James D. Malley

Statistical Applications of Jordan Algebras

Springer-Verlag
New York Berlin Heidelberg London Paris
Tokyo Hong Kong Barcelona Budapest

James D. Malley
National Institutes of Health
Division of Computer Research and Technology
Building 12A, Room 3053
Bethesda, Maryland 20892

Library of Congress Cataloging-in-Publication Data Available
Printed on acid-free paper.

© 1994 Springer-Verlag New York, Inc.
Softcover reprint of the hardcover 1st edition 1994

Camera ready copy provided by the author.

9 8 7 6 5 4 3 2 1

ISBN-13: 978-0-387-94341-1   e-ISBN-13: 978-1-4612-2678-9
DOI: 10.1007/978-1-4612-2678-9

Preface

This monograph brings together my work in mathematical statistics as I have viewed it through the lens of Jordan algebras.

Three technical domains are to be seen: applications to random quadratic forms (sums of squares), the investigation of algebraic simplifications of maximum likelihood estimation of patterned covariance matrices, and a more wide-open mathematical exploration of the algebraic arena from which I have drawn the results used in the statistical problems just mentioned. Chapters 1, 2, and 4 present the statistical outcomes I have developed using the algebraic results that appear, for the most part, in Chapter 3.

As a less daunting, yet quite efficient, point of entry into this material, one avoiding most of the abstract algebraic issues, the reader may use the first half of Chapter 4. Here I present a streamlined, but still fully rigorous, definition of a Jordan algebra (as it is used in that chapter) and its essential properties. These facts are then immediately applied to simplifying the M–step of the EM algorithm for multivariate normal covariance matrix estimation, in the presence of linear constraints, and data missing completely at random. The results presented essentially resolve a practical statistical quest begun by Rubin and Szatrowski [1982], and continued, sometimes implicitly, by many others. After this, one could then return to Chapters 1 and 2 to see how I have attempted to generalize the work of Cochran, Rao, Mitra, and others, on important and useful properties of sums of squares.

It is in Chapter 3 that I have gone most thoroughly into the underlying, industrial-strength mathematical details, with the intent of pushing ahead the theoretical frontier in order that paths might be laid out to facilitate still other statistical applications of Jordan algebras. Most of these results originally appeared in Malley [1987], and represent my algebraic response to a need to solve certain specifically statistical questions that have preoccupied me. It has been my experience that there is much to be gained for both the statistical and the mathematical communities by this intellectual boundary crossing and intertwining of goals and ideas. I hope that some readers will also come to enjoy this process of discovering common ground in unlikely places.

Finally, I want to thank a number of friends who have helped in matters technical and non-technical: Kevin McCrimmon, Nathan Jacobson, Ted Szatrowski, and Linda Hirsch, come first to mind. And my uncommonly loyal editor at Springer-Verlag, Martin Gilchrist, has been a steady, positive resource on this project during its multi-year fabrication. I want to express my appreciation for the considerable and generous help of my wife Karen: while this work has been long in development, her support has been unconditional. Lastly, this book is for you Dean, now out there exploring new territory for all of us.

James D. Malley
Bethesda, Maryland

Table of Contents

Table of Contents

*When we try to pick out anything by itself, we find
it hitched to everything else in the Universe.*
John Muir

Chapter 1   Introduction

In statistics it frequently occurs that we need to know whether two sums of squares are independent and whether they are each chi-square in distribution. In terms of necessary and sufficient conditions, the answers to these questions have been worked out, by Cochran, Craig, Rao, Mitra and others, over a period of several decades. For multivariate normal data, these results are usually expressed in terms of rather complex equations involving the true data mean vector and the true covariance matrix (assumed known at least up to a constant).

A natural generalization of these results allows for the covariance matrix for the data to be not known exactly, but known to be contained in a given linear space of matrices. And rather than this increased level of generality then making the necessary and sufficient conditions still more involved, it is possible, using methods of advanced algebra, to produce conditions that are actually less complex, easier to apply and transparently intuitive.

This is one of the primary results we obtain in Chapter 2. In language often used by mathematicians when describing the solution to a problem, we go "upstairs," solve an apparently harder or different problem and then apply the solution to the original question. It has often been found that the generalized problem is much easier to solve, hence this problem solving scheme can be quite powerful, provided the suitable, simplifying generalization can be found.

In another direction, the EM algorithm for maximum likelihood estimation has earned a widespread and well-deserved popularity, and indeed history discloses that it has been in use for some time in other guises, under other names (see Dempster, Laird and Rubin [1977]). Yet an important computational simplification of the algorithm for covariance matrix estimation depends on an abstract algebra condition. In particular, for multivariate normal data we show it is always possible, using this condition, to find maximum likelihood (ML) estimates for the covariance matrix via the EM algorithm, in a procedure for which the M–step effectively vanishes, and convergence is guaranteed whenever the likelihood is unimodal.

A simple, again algebraic, extension of our method applies as well to any multinormal data covariance matrix estimation problem where data may be missing completely at random.

The core strategy underlying the problems and solutions just discussed is the central theme of this monograph: many interesting and useful statistical results can be derived or at least better understood once they are located in an appropriate algebraic context. Our focus here will be on showing how the abstract objects called **Jordan algebras** provide one such context, for the problems discussed above and many others as well.

Indeed, two of these other problems have solutions that have either been in the literature for some time and are still in use, or, are quite recent: we show that both are in fact effective procedures in part because they share an implicit use of Jordan algebras; full details appear in Chapter 4. The two solutions we have in mind here are, respectively, the restricted maximum likelihood (REML) method for variance component estimation, and a recent significant computational advance for ML covariance estimation for stationary times series.

More precisely, it has been known for some time that REML is, in effect, the EM algorithm, for which the M–step is completely trivial (see Dempster, Laird and Rubin [1977; p.17]). On the other hand, Dembo at al. [1989] show how to obtain ML estimates of a stationary time series covariance matrix, again using a form of the EM algorithm with a greatly streamlined M–step. In both of these cases, the simplifying use of Jordan algebras was implicit, and not noted as such. Our observation about these methods is that, after the technical dust settles, the key M–step simplifications can be seen to hold precisely because a certain vector space of real symmetric matrices forms a Jordan algebra. Evidently the validation of a Jordan algebra approach to these statistical problems is being made here for the first time.

In still another direction, a long-standing problem connected with the actual invention of Jordan algebras (in 1934) can be partially put to rest.

Thus while Jordan algebras were originally devised to axiomatize quantum mechanics (QM), they were eventually found to be technically inadequate for this difficult program, and did not represent a distinctive or especially useful starting point. (But here see Upmeier [1987] who puts a more positive spin (pun intended) on this issue.)

However, it can be shown that narrowing the algebraic agenda in the search for rational, simple axioms for QM can return Jordan algebras to a place of importance in QM. Namely, they return for the study of optimal statistical inference for quantum mechanical data. Briefly, it can be shown that an important class of statistical decision operators forms a Jordan algebra. Full details appear elsewhere, since the required preliminaries and surrounding mathematical and physical issues require their own monograph; for the present, see Malley and Hornstein [1993; preprint 1992].

Further motivation for an algebraic approach to a wide variety of statistical problems, is gained by survey of another historical perspective.

That is, for many years the intertwined areas of experimental design and combinatorics have benefitted from the use of advanced algebra, for example Galois fields and association schemes. Also, Speed [1987] and his co-workers, along with the complementing, alternative work of Dawid [1988], have recently invoked associative algebras and methods of group theory to facilitate understanding of what really constitutes an analysis of variance. Concerning the techniques employed by Dawid [1988], statistics has already for some time employed elegant and powerful methods that depend on group structure and group invariance; see for example Diaconis [1988], Eaton [1989], and Wiisman [1990].

Closer in spirit to our work here has been the research of James [1957], Mann [1960], Bose [1959] and others who produced algebraic contexts for unifying and studying linear models and the associated hypothesis tests. They all made use of associative algebras, one of the three main sequence algebras, the other two being the Lie and Jordan algebras. These three classes are quite intimately connected, so much so that results for one of them are often used in proofs dealing with the others.

However, one gleaning from this statistical history is that, evidently, only the associative algebras have been extensively or explicitly applied. For example, the study of zonal polynomials has contributed much to multivariate distribution theory, and associative algebra techniques have been used to elegant advantage in this area; see for example Farrell [1985]. Moreover, zonal polynomials also have a rich, independent Lie algebra life in the context of "invariant differential operators"; see Helgason, [1985].

So, Lie algebra notions may be found in and around many important statistical areas, but to our knowledge have not yet directly surfaced in the statistical literature. We expect this to change with time, given the ongoing, cooperative commingling of advanced algebraic ideas and mathematical sta-tistics.

In contrast, Jordan algebras have evidently been first formally introduced to statistical readers only fairly recently, in the area of optimal variance com-ponent estimation and the work of Seely [1971], Pukelsheim [1981] and others; see also Malley [1986]. We believe a more extensive role exists for them: applied and theoretical statistics can be advanced in important and useful directions using these relatively abstract structures.

At this point, some first definitions and an example might help.

An **algebra** is, basically, a vector space with a multiplication. Familiar examples include the real and complex numbers and functions, and matrices over the real or complex numbers.

Now, let A be a square matrix with real number entries. This much-used object of statistics can be naturally decomposed into other matrices over the reals, and these matrices belong to classes that in turn have an independent, highly developed structure of their own. Thus A can always be written as:

$$A = \tfrac{1}{2}(A + A^T) + \tfrac{1}{2}(A - A^T)$$

with A^T the transpose of A.

The first term on the right is a basic element of a Jordan algebra, being closed under addition (sums of symmetrics are again symmetric), and closed under multiplication using the Jordan product:

$$A.B = \tfrac{1}{2}(AB + BA)$$

(The coefficient $\tfrac{1}{2}$ is put in to make $A . A = A^2$.)

The second term on the right in our decomposition for A is an element of a Lie algebra, being closed under addition (sums of skew-symmetrics are again skew-symmetric), and closed under multiplication using the Lie product:

$[A, B] = AB - BA.$

(The coefficient $\frac{1}{2}$ is not needed now since $[A, A] = 0 = 2[A, A]$.)

Another basic Lie example is the space of real 3–vectors which add as usual, but which use the vector space cross product.

The decomposition of A above can be carried a step further, in that the standard associative matrix product also splits into a Jordan and a Lie part:

$$AB = A.B + \tfrac{1}{2}[A, B].$$

These two spaces, the symmetric and the skew-symmetric matrices, are the canonical examples of Jordan and Lie algebras.

A consequence to be drawn from above is that both Jordan and Lie algebras are visible as primitive parts of familiar objects in statistics. It is by focusing on the Jordan part that we propose, in the following pages, to fully resolve or at least illuminate some important statistical questions.

Two initial instances of this will be a thorough study of the examples first mentioned in this introduction: sums of squares and hypothesis tests in the mixed linear model (Chapter 1).

Continuing with an outline of the text, Chapter 3 is quite technical, containing several new algebraic results that in turn lead to complete proofs for the results used in Chapter 2. Chapter 4 is a mix of algebra, some new, some old, and applications to ML estimation with the EM algorithm. Most of these applications are new, but a few have the different purpose of showing how some known statistical methods in fact rely on Jordan algebra methods.

We note that Jordan methods also play a pivotal role in a discussion of nuisance parameters that may be introduced in some of our implementations of the EM algorithm. This will invoke use of the differential geometric approach to statistical inference, and it is interesting, but perhaps not so surprising in light of the connections displayed above, that Lie algebras also make a brief appearance here.

Finally, at many points in this monograph a cyclic plan of application, theory, application,..., is used in an effort to keep the statistical results from being orphaned in a thick fog of abstract algebra. The sequence is also an artifact of the way in which this work evolved, with many of the Jordan results arising as new algebraic facts each derived to answer a specific statistical question, and conversely. It is hoped that this pattern also serves to display an underlying unity and a commonality of purpose, and so generating new questions, results and techniques that may be of interest to both the statistical and the mathematical communities.

About notation: completions of proofs are denoted by ■, while conclusions of **Remarks** or **Examples** are denoted by □. Lemmas, theorems and corollaries are numbered in each chapter, as for example, **Theorem 5.2**, that names the second theorem in section five of a chapter. References are given separately at the end of each chapter, with the sole exception of this first chapter: these are rolled into those appearing at the end of Chapter 2.

Chapter 2   Jordan Algebras and the Mixed General Linear Model

2.1 Introduction.

Jordan algebras are not commonly a standard technical tool for even many researchers in pure algebra. It is important therefore to begin by defining and illustrating them in a statistically accessible context. Hence the required definitions, theorems and proofs will be introduced with the minimum of mathematical embellishment. Some of the abstract algebra thus set aside will then appear as **Remarks**, some deferred to Chapter 3, and some even further removed to literature citations.

Wherever feasible several equivalent definitions will be given, and alternative proofs presented. The reader is encouraged to keep at hand a small set of low–order symmetric matrices with which to test and probe the content of the definitions and theorems.

With only an occasional exception, notation in this chapter will be consonant with that used by algebra researchers: sets and spaces will be written using capitals, elements of these sets and spaces appearing as lower case. Thus a $\in$ A will mean: a is an element of the set (or space) A. Also, matrices will be written as lower case, except as noted. Our rationale for this choice is that many of the equations to appear will be of the form aSb = 0, where a and b are *single* elements, and S is a *collection* of elements: we consider this better than writing pages of equations set in bold face, such as: **ASB = 0**. (We consistently make one exception to this rule, in that we routinely use **I** to denote the identity matrix.)

Our choice is not the conventional one for statistics, but it is hoped that this shift asked of the reader is relatively uneventful, at least easier on the eyes, and that it ultimately helps to make reading the original abstract algebra source literature less difficult. Consonant with the rationale of this choice, we will in Chapter 4 use notation in line with that of the EM literature, and again, it is hoped that our context-dependent notation is a help and not a hindrance.

2.2 Square Matrices and Jordan Algebras.

Let's begin again, as in Chapter 1, by considering a decomposition for any real matrix. The following will be standard notation:

$\Re$ = real numbers,
$[\Re]_m$ = real m × m matrices,
S_m = real m × m symmetric matrices.

Then for any $a \in [\mathfrak{R}]_m$:

$$a = \tfrac{1}{2}(a + a^T) + \tfrac{1}{2}(a - a^T),$$

for a^T the transpose of matrix a. Observe that the first term in the sum is always symmetric,

$$b = \tfrac{1}{2}(a + a^T) \in \mathbf{S}_m,$$

while the second term is always skew-symmetric:

$$b = \tfrac{1}{2}(a - a^T) \quad \text{implies} \quad b^T = -b.$$

The decomposition provides a key connecting link between the three most important (classical) "algebras" in mathematics, where by **algebra** we mean: a vector space that also has a well-defined multiplication operation, one obeying the usual **distributivity** rules. Thus if, for the moment, we write this multiplication as " $*$ " then the operation must satisfy:

$$\lambda(a + b) = \lambda a + \lambda b, \qquad\qquad \lambda(a * b) = (\lambda a) * b = a * (\lambda b)$$
$$a * (b + c) = a * b + a * c \qquad\qquad (b + c) * a = b * a + c * a$$

for λ any real or complex number.

These conditions define a real (or complex) algebra if all the coefficients λ are real (or complex). Latter, other kinds of coefficients will appear: we'll use algebras over fields other than the reals or complexes.

The **dimension of an algebra** A will be its dimension as a vector space, written: dim A .

Note that it is *not* explicitly required that the algebra multiplication to be

   commutative: $a * b = b * a$,

or that it be

   associative: $(a * b) * c = a * (b * c)$.

Some of the algebras used below will have one, or the other, or both of these additional properties. Familiar examples that do have both are real and complex numbers and real- or complex-valued functions. Note that the complex numbers can be considered as a real or complex algebra: the coefficients in the calculations of interest may be fixed as all real or complex.

A familiar associative, but not commutative algebra is $[\mathfrak{R}]_m$, for then:

$$a(bc) = (ab)c, \qquad \text{for all } a, b, c \in [\mathfrak{R}]_m,$$

but in general ab $\neq$ ba.

We come now to the definition of that algebra which is central to this work. A **Jordan algebra** is an algebra (= vector space with multiplication) such that it's product " . " obeys the two rules:

J1: $a . b = b . a$
J2: $a^2 . (b . a) = (a^2 . b) . a$ where $a^2 = a . a$

These rules, especially **J2** may well appear cryptic and unmotivated. However, more practical, transparent and equivalent definitions of a Jordan algebra will be given shortly, since only a certain type of Jordan algebra will be used in this work. For now just note that the algebra is commutative (this is **J1**), but has only a most restricted kind of associativity (given by **J2**).

Again, the fundamental example of a Jordan algebra is $\mathbf{S}_m$, the space of real symmetric matrices of order m. The Jordan product is defined as:

$a . b = \frac{1}{2}(ab + ba)$.

Note that $\dim \mathbf{S}_m = \frac{1}{2}m(m + 1)$

The symmetrized product $a . b$, just defined, will not in general return a symmetric matrix if a or b is not symmetric. Also observe that the algebra and its product are defined using the elements and ambient multiplication of something already known, the matrices $[\mathfrak{R}]_m$. This is a frequently–used technical device in algebra for building new objects from old, but some care is usually needed to make such constructions unambiguous. Hence we state:

In all that follows, the unadorned product ab will mean the usual matrix product for arbitrary elements a and b $\in [\mathfrak{R}]_m$.

Remark. A Jordan algebra need not be (algebra) isomorphic to a subalgebra of $\mathbf{S}_m$, with the product " . " given as above. It is standard to call those that do have this property: **special** Jordan algebras. A complete characterization of an important class of such algebras will be given in Chapter 3. It is not an easy result to prove, although the outcome is the intuitive one: by adding one qualification to the class of all special Jordan algebras to be studied, it is found that only a small family of well-defined, distinct collections in $\mathbf{S}_m$ is possible.

Finally, observe that for any element a of a special Jordan algebra:

$a^2 = a . a = \frac{1}{2}(a^2 + a^2) = a^2$. $\square$

Examples.
   (1) In $\mathbf{S}_2$ let $\mathscr{S}$ be the real vector space spanned by $\mathbf{I}_2$ and the matrix

$a = \begin{pmatrix} 1 & 2 \\ 2 & 1 \end{pmatrix}$.

$\mathscr{S}$ is in fact the space of all matrices s of the form

$$s = \begin{pmatrix} \alpha & \beta \\ \beta & \alpha \end{pmatrix}$$

and it can also be checked, albeit tediously, that $\mathscr{S}$ is a Jordan algebra.

(2) In $\mathbf{S_3}$ let $\mathscr{S}$ be the space spanned by $\mathbf{I_3}$ and the matrix

$$a = \begin{pmatrix} 1 & 2 & 3 \\ 2 & 1 & 2 \\ 3 & 2 & 1 \end{pmatrix}.$$

$\mathscr{S}$ is now the space spanned by all matrices s of the form

$$s = \begin{pmatrix} \alpha + \beta & 2\alpha & 3\alpha \\ 2\alpha & \alpha + \beta & 2\alpha \\ 3\alpha & 2\alpha & \alpha + \beta \end{pmatrix}$$

and is **not** Jordan: **J1** holds, **J2** does not.

In both the examples above, and generally, checking to see if a space $\mathscr{S}$ is Jordan can be tedious. The equivalent definitions for a Jordan algebra given below will simplify this. A hint of things to come: in each example check to see if an arbitrary $a \in \mathscr{S}$ is such that $a^2 \in \mathscr{S}$; what if a has at most two eigenvalues? □

Remark. The skew-symmetric matrices form a key example of a Lie algebra, as noted earlier. Technical facts about Lie algebras will be not required in this work, but for completeness, a **Lie algebra** is defined to be: a vector space with multiplication " $[\,.\,,\,.\,]$ " such that:

L1: $[a,\,a] = 0$
L2: $[[a,\,b],\,c] + [[b,\,c],\,a] + [[a,\,c],\,b] = 0$.

Observe that **L1** implies $[a,\,b] = -\,[b,\,a]$, and so is said to be **anticommutative**. The reader might find it useful here to check **L1**, and **L2** for: (i) the real vectors of size three, using the vector space cross product, or, (ii) the skew-symmetric matrices. Once more note that checking this algebra's replacement for associativity, **L2**, is the most tedious step. □

2.3 Idempotent and Identity Elements.

We proceed to build some technique with things Jordan, beginning with a look at the behavior of some useful properties of single elements.

Definition. An element e of any subspace $\mathscr{S}$ contained in $[\mathfrak{R}]_m$, is an **associative identity element** if $es = se = s$, for all $s \in \mathscr{S}$.

Remark. For an arbitrary proper subspace $\mathcal{S}$ of $[\mathfrak{R}]_m$, the element e *cannot* be presumed to be the identity matrix $\mathbf{I}_m \in [\mathfrak{R}]_m$. As an example consider the following mixed model, where we briefly revert to standard statistical notation for matrices.

Let the m-dimensional data vector $\mathbf{y}$ have mean $\mathbf{X}\alpha$ and covariance matrix Σ. Assume rank $\mathbf{X} < m$, and that Σ is positive definite and of the form

$$\Sigma = \sum \lambda_i G_i$$

for $\mathbf{G}_i$ some finite set in $\mathbf{S}_m$, and some set of $\lambda_i \in \mathfrak{R}$. Let $\mathcal{S}$ be the space in $\mathbf{S}_m$ spanned by the $\mathbf{G}_i$. Put

$$\mathbf{M} = \mathbf{I}_m - \mathbf{X}^T(\mathbf{X}^T\mathbf{X})^+\mathbf{X}^T,$$

for $(\mathbf{X}\mathbf{X}^T)^+$ the Moore–Penrose g–inverse of $(\mathbf{X}\mathbf{X}^T)$. $\mathbf{M}$ is the projection onto the orthogonal complement of the column space of $\mathbf{X}$. Write $\mathbf{y}^* = \mathbf{M}\,\mathbf{y}$. Then the transformed data vector $\mathbf{y}^*$ has mean zero, and positive semidefinite covari-ance matrix

$$\Sigma^* = \mathbf{M}\Sigma\mathbf{M}.$$

Σ^* is an element of the space $\mathcal{S}^*$ of matrices of the form $\mathbf{M}s\mathbf{M}$ for all $s \in \mathcal{S}$.

We find: the space $\mathcal{S}^*$ has a new (associative) identity element $\mathbf{M}$, and $\mathbf{I}_m \notin \mathcal{S}^*$. The example is both "pre-Jordan" and even "pre-statistical," since it is just an observation about spaces of matrices. ◻

Continuing with the discussion of identity elements, a Jordan identity element $e \in \mathcal{S}$ would then, in a parallel manner, be such that $e \cdot s = s$, for all $s \in \mathcal{S}$. However the two kinds of identity elements, associative and Jordan, are really the same, by a standard result. This is verified by proving something slightly more general, and needed for this is:

Definition.   An element $e \in [\mathfrak{R}]_m$ such that $e^2 = e$ is said to be **idempotent**.

Recall that in a special Jordan algebra $e \cdot e = e^{2\cdot} = e^2$, so that Jordan and associative idempotents are the same.

3.1 Lemma. Suppose $\mathcal{S}$ is a space in $\mathbf{S}_m$, and $e \in \mathcal{S}$ is an idempotent. If $e \cdot s = s$ for some $s \in \mathcal{S}$, then also $e \cdot s = es = se = s$.

Proof. In any special Jordan algebra the following can, with a little effort, be verified:

$$aba = 2a \cdot (a \cdot b) - a^2 \cdot b.$$

[Just write out all the products, using $a \cdot b = \tfrac{1}{2}(ab + ba)$.] Since e is idempotent

$$ese = 2e \cdot (e \cdot s) - e^2 \cdot s = 2s - s = s,$$

so that

$$es = e(ese) = ese = s = e \cdot s.$$

The last equality follows from the assumption that $e \cdot s = s$. Verifying that

$$se = s = e \cdot s$$

is similar to the above, and this completes the proof. ∎

From **Lemma 3.1** it follows immediately that a Jordan identity element of a space $\mathscr{S}$ is also an associative identity element, and conversely. Also true, and equally standard, is that idempotents are **Jordan orthogonal** if and only if they are **associative orthogonal**. The needed definitions and results are contained in:

3.2 Lemma.

(i)   If $e \cdot s = s$ for all $s \in \mathscr{S}$ then $es = s = se$, and conversely;

(ii)   In special Jordan algebras, idempotents are Jordan orthogonal if and only if they are associative orthogonal: for idempotents e_1, e_2 we have
$$e_1 \cdot e_2 = 0 \quad \Leftrightarrow \quad e_1 e_2 = 0;$$

(iii) Subspaces $\mathscr{S}$ and $\mathscr{T}$ of $\mathbf{S}_m$, with identities $s_0 \in \mathscr{S}$, $t_0 \in \mathscr{T}$, are Jordan orthogonal; that is $\mathscr{S} \cdot \mathscr{T} = 0$: $s \cdot t = 0$ for $s \in \mathscr{S}$, $t \in \mathscr{T}$, if and only if they are associative orthogonal (i.e. $\mathscr{S}\mathscr{T} = \mathscr{T}\mathscr{S} = 0$: $st = ts = 0$, for all $s \in \mathscr{S}$, $t \in \mathscr{T}$).

Proof. Part **(i)** is immediate from **Lemma 3.1**, and for Part **(ii)** consider idempotents a, b with $a \cdot b = 0$; the converse for Part **(ii)** is clear. Now

$$aba = \tfrac{1}{2} a(a \cdot b)a = 0,$$

so that

$$ba = 2(b \cdot a)a - aba = 0,$$

and similarly for ab.

For Part **(iii)** we need only show that Jordan orthogonal implies associative orthogonal, since the converse is clear. Thus for

$$\begin{aligned} 2(s \cdot t) &= 2(s \cdot s_1) \cdot (t_1 \cdot t) \\ &= 2(s_1 s s_1) \cdot (t_1 t t_1) \end{aligned}$$

using Part **(i)**,

$$= s_1 s s_1 t_1 t t_1 + t_1 t t_1 s_1 s s_1.$$

Since $s_0^2 = s_1$, $t_1^2 = t_1$, and $s_1 \cdot t_1 = 0$, it is implied by Part **(ii)** that

$$s_1 \cdot t_1 = t_1 \cdot s_1 = 0,$$

so that $s \cdot t = 0$ as required, and the proof is complete. ∎

2.4 Equivalent Definitions for a Jordan Algebra.

This section develops a set of equivalent, practical definitions for a sub-space of $\mathbf{S}_m$ to be a Jordan algebra. The first is standard in the Jordan algebra literature, and evidently first appeared in the statistical literature in Seely [1971]. In applications, keep in mind a key premise of the lemma, that the space contains an identity element, and also recall that this element need not be the identity matrix.

4.1 Lemma. Let $\mathscr{S}$ be a subspace of $\mathbf{S}_m$. Suppose that $\mathscr{S}$ contains an identity element e. Then $\mathscr{S}$ is a Jordan algebra if and only if any one of the following

equivalent conditions holds:
  (i) $ab + ba \in \mathscr{S}$, for all a and $b \in \mathscr{S}$;
  (ii) $aba \in \mathscr{S}$, for all a and $b \in \mathscr{S}$;
  (iii) $a^2 \in \mathscr{S}$, for all $a \in \mathscr{S}$.

Proof. Let $c = a - e$, so $c \in \mathscr{S}$. Then
  $$aba = (c + e)b(c + e) = c^2 + b + (cb + bc).$$
Given (i) it follows that
  $$c^2 = \tfrac{1}{2}(c^2 + c^2) \in \mathscr{S},$$
and also that
  $$2c \cdot b = cb + bc \in \mathscr{S}.$$
Hence $aba \in \mathscr{S}$, and this shows that (i) implies (ii).

Next, given (ii) put $b = e$ to get (iii) immediately.

Finally, suppose (iii) holds and put $c = a + b$. Then
  $$c^2 = (a + b)^2 = a^2 + b^2 + (ab + ba) \in \mathscr{S},$$
so $(ab + ba)$ and also
  $$a \cdot b = \tfrac{1}{2}(ab + ba) \in \mathscr{S},$$
and this last yields (i). This completes the proof. ■

Remark. Note that $a^2 \in \mathscr{S}$ and $aba \in \mathscr{S}$ implies that $a^n \in \mathscr{S}$ for all $n \geq 1$. Thus any $\mathscr{S}$ that is Jordan contains all real polynomials in any single element $a \in \mathscr{S}$. ☐

Appearing as Lemma 1 of Jensen [1988] is another characterization of Jordan algebras of $\mathbf{S}_m$. We proceed to give alternative versions of this result, along with a new proof of Jensen's lemma. In one direction the results are essentially particular cases of a result of McCrimmon [1969]. In the other direction they are immediate consequences of a basic result given in Jacobson [1968; p. 2]. Observe now that the space $\mathscr{S}$ in question must actually contain the identity matrix $\mathbf{I}$, not just an identity element e.

4.2 Lemma. Suppose $\mathscr{S}$ is a finite dimensional vector space contained in $\mathbf{S}_m$, and that $\mathscr{S}$ contains the identity matrix $\mathbf{I}$. Then:
  (i) $\mathscr{S}$ is a special Jordan algebra if and only if it contains the Moore–Penrose inverse a^+ of every element $a \in \mathscr{S}$;
  (ii) $\mathscr{S}$ contains all its true inverses s^{-1}, for $s \in \mathscr{S}$, if and only if it contains all its Moore–Penrose inverses;
  (iii) $\mathscr{S}$ contains all its true inverses s^{-1}, for $s \in \mathscr{S}$, if and only if it contains the inverse of every positive definite element in $\mathscr{S}$.

Proof. Part (ii) can be seen to follow directly from (i), while the proof of (i) is effectively also a proof of (iii). Thus we need only prove (i).

If $\mathscr{S}$ contains all its Moore–Penrose inverses a^+ then it must also contain all its true inverses a^{-1}. To show that this implies $\mathscr{S}$ is Jordan, a standard, nonetheless remarkable, identity is employed. It is derived here for the convenience of the reader; see Jacobson [1968; p. 2].

Thus, suppose a and b are invertible, in $\mathscr{S}$, and $a \neq b^{-1}$. Then

$$a^{-1} + (b^{-1} - a)^{-1} = a^{-1}[(b^{-1} - a) + a](b^{-1} - a)^{-1}$$

$$= a^{-1}b^{-1}(b^{-1} - a)^{-1}$$

$$= [(b^{-1} - a)ba]^{-1} = [a - aba]^{-1}$$

from which **Hua's identity** follows:

$$a - (a^{-1} + (b^{-1} - a)^{-1})^{-1} = aba.$$

Next, recall the basic inequality involving any matrix $c \in \mathbf{S_m}$:

$$\max \; [(x^T cx) / (x^T x)] = \lambda_{max},$$

where the maximum is taken over all $x \in \mathfrak{R}_m$, $x \neq 0$, and λ_{max} is the largest eigenvalue of the matrix c. Using the inequality there always exists a sufficiently small, positive constant ζ, such that $I - \zeta c$ is positive definite. Choose such a ζ, fix it and next find a sufficiently small, positive constant t such that
$$I - t(I - \zeta c)$$
is also positive definite.

In Hua's identity use $b = I$, and $a = t(I - \zeta c)$. The matrix a is positive definite since $t > 0$, and $(I - \zeta c)$ is positive definite. By our assumption on $\mathscr{S}$, that $I \in \mathscr{S}$, we get:

$$a - (a^{-1} + (I - a)^{-1})^{-1} \in \mathscr{S}.$$

so that

$$a^2 = a(I)a$$

$$= t^2 (I - 2\zeta c + \zeta^2 c^2) \in \mathscr{S}.$$

This implies $c^2 \in \mathscr{S}$, for all $c \in \mathscr{S}$, and this direction of the proof is complete by **Lemma 4.1(iii)**.

The converse, that every Jordan algebra of real symmetric matrices must contain all its Moore–Penrose inverses, is given in Malley [1986; p. 103]. There it is pointed out that the Lagrange-Sylvester interpolation formula for a^+ (see Rao and Mitra [1971; p. 62]) allows one to write a^+ as a polynomial in a. That is:

$$a^+ = (a^2)^+ a$$

$$= \Sigma \, (\alpha_i)^{-2} (g_i / h_i) a,$$

for $\{\alpha_i\}$ the set of distinct non-zero eigenvalues of a, and

$$g_i = \prod_{j \neq i}\left[a^2 - (\alpha_j)^2 \mathbf{I}\right]$$
$$h_i = \prod_{j \neq i}\left[(\alpha_i)^2 - (\alpha_j)^2\right].$$

Consequently $a^+ \in \mathscr{S}$ and the proof is complete. ∎

Remark. Lemma 1 of Jensen [1988] is **3.4(iii)** above (proved here by a different method). ☐

Remark. The converse part of **Lemma 4.2(i)** above can be considerably extended. Thus call a Jordan algebra $\mathscr{J}$ **non-degenerate** if for $u \in \mathscr{J}$, we have: uau = 0 for all $a \in \mathscr{J}$ if and only if u = 0. Next, call an element $u \in \mathscr{J}$ **regular** if there exists $y \in \mathscr{J}$ such that u = uyu. Observe that this generalizes the notion of a g–inverse to non-matrix settings. Then, as a special case of a result of McCrimmon [1969], it follows that every finite dimensional, non-degenerate Jordan algebra $\mathscr{J}$ is such that every element $u \in \mathscr{J}$ is regular. ☐

Now, the statistical questions studied here will use properties of the finite dimensional Jordan algebras contained in $\mathbf{S_m}$. Malley [1987], and Jacobson [1987] (c.f. Jensen [1988]) were able to fully characterize all these algebras, and this classification is ultimately shown to be central to understanding independence and χ^2–distributivity for random quadratic forms in normally distributed variables. Full details appear below in Chapter 3; a full statement (only) of the result is given here.

Let's start with a basic fact often used later:

4.3 Lemma. $\quad tr(aa^T) = \Sigma(a_{ij})^2 = 0 \quad \Leftrightarrow \quad a = 0, \quad$ for any $a \in [\mathfrak{R}]_m$.

Remark. From **Lemma 4.3** it follows immediately that subalgebras of $\mathbf{S_m}$ are **formally real**, in the sense that:

$$a^2 + b^2 = 0 \quad \Leftrightarrow \quad a = b = 0, \quad \text{for all a, b} \in \mathbf{S_m}.$$

Formal reality of certain Jordan subalgebras is a key fact in the most complete algebraic statement of the classification given below, but one that can always assumed without comment for algebras in $\mathbf{S_m} \subseteq [\mathfrak{R}]_m$. ☐

Next, an **ideal** $\mathscr{C}$ of any algebra $\mathscr{A}$ (not necessarily), is defined to be a subspace of $\mathscr{A}$ which is also a subalgebra (of the same type, Lie or Jordan, etc.) and is such that

$a * c$, and $c * a \in \mathscr{C}$ for all $a \in \mathscr{A}$, $c \in \mathscr{C}$, using the algebra product " $*$ ". Call an algebra $\mathscr{A}$ **simple** if $\mathscr{A}^2 = \mathscr{A} * \mathscr{A}$, its product with itself, is not zero, and if it does not contain any proper, non-zero ideals. $\mathbf{S_m}$ is the canonical example a simple Jordan algebra, while $[\mathfrak{R}]_m$ is a simple associative algebra.

Remark.   For associative and Lie algebras, the product of two ideals $\mathscr{C}$ and $\mathscr{D}$ that is, the linear space spanned by all products of pairs of elements of $\mathscr{C}$ and $\mathscr{D}$, is easily shown to always be another ideal of the algebra. That this is not the case, however, for Jordan algebras is a non-trivial fact to be reckoned with in many Jordan calculations. $\square$

Remark.   The space of all trace zero matrices in $[\mathfrak{R}]_m$, $m \geq 2$, is the simple classical Lie algebra **sl(m)** (see e.g. Sagle and Walde [1973; p. 61]). The space of skew–symmetric matrices is the Lie algebra of the special orthogonal Lie group **so(m)** of all orthogonal matrices with determinant one. The skew–symmetrics form a simple Lie algebra, but not a Lie ideal in the space of trace–zero matrices. $\square$

Finally, let's agree to call two matrix subspaces $\mathscr{S}_1$ and $\mathscr{S}_2 \subseteq [\mathfrak{R}]_m$, **orthogonally equivalent** if there is an orthogonal matrix u such that:

$$\mathscr{S}_1 = u\mathscr{S}_2 u^T$$

The basic Jordan classification result can now be stated (complete details appear in Chapter 3):

4.4 Theorem   (Jacobson [1987]; c.f. Malley [1986], Jensen [1988]).
Every special, formally real, finite dimensional Jordan algebra is isomorphic to a subalgebra of $\mathbf{S_m}$, for some m. Every such Jordan algebra is ortho-gonally equivalent to a direct sum of simple Jordan ideals, and has an identity element e, with e . a = a for all a $\in \mathscr{J}$. Each of the simple Jordan ideals is algebra isomorphic and orthogonally equivalent to one of the following, where each case is realized for some class of real symmetric matrices:

  (i) $\mathbf{S_m}$, real n × n symmetric matrices, for some n;
  (ii) $\mathscr{H}([C]_n$, *) = the class of n × n (n $\geq$ 3) matrices over C (the complex numbers) left fixed by (·)*, the complex conjugate transpose;
  (iii) $\mathscr{H}([Q]_n$, *) = the class of n × n (n $\geq$ 3) matrices over Hamilton's quaternion algebra Q left fixed by the usual quaternion conjugation followed by matrix transpose;
  (iv)   The Jordan algebra $\mathscr{J}(\mathscr{V}, f)$ of a positive definite inner product f(x, y) on a real vector space $\mathscr{V}$, $\dim \mathscr{V} \geq 2$. An element of $\mathscr{J}(\mathscr{V}, f)$ has the form a = (α1 + x), for $\alpha \in \mathfrak{R}$, x $\in \mathscr{V}$, and has product defined by:
    $(\alpha 1 + x) . (\beta 1 + y) = [\alpha\beta + f(x,y)]1 + [\beta x + \alpha y]$.

Orthogonal equivalence in **Theorem 4.4** means, more precisely, that there is an orthogonal matrix u that is also block diagonal:

$$u = \text{diag}\,[u_1, \ldots, u_k]\,, \quad \text{each } u_i \text{ orthogonal,}$$

so that after relabeling:

$$u\,\mathscr{J}\,u^T = \text{diag}\,[\mathscr{J}_1, \ldots, \mathscr{J}_k]\,,$$

with

$$\text{diag}\,[0, \ldots, \mathscr{J}_i, \ldots, 0]$$

being the ith simple component of $\mathscr{J}$, given by one of the cases in **4.4**.

Remark. The important fact that the decomposition can be obtained by an orthogonal change of basis appears in Jensen [1988]. It also is given in Jacobson [1987], who obtained it along the way to solving the more difficult problem of finding all the orbits under the orthogonal group of the Jordan algebras of real symmetric matrices. ❑

In order to connect the above abstract algebra with our original statistical context, consider next the relationships between arbitrary sets of real symmetric matrices and certain Jordan algebras derived from them.

2.5 Jordan Algebras Derived from Real Symmetric Matrices.

Begin by considering an arbitrary set of matrices contained in S_m. Since S_m is finite dimensional over $\mathfrak{R}$, so are all its subspaces, and thus there is some finite set $\{m_i\}$ that forms a basis for the vector space spanned by the given arbitrary set in S_m. To avoid having to write this out every time we use:

Definition.   For any set $\{a_i\}$ of matrices in $\mathbf{S_m}$, write $\mathscr{S}(\{a_i\})$ for the vector space spanned by the set $\{a_i\}$.

Remark. If $\mathbf{I} \in \mathscr{S}$ then using the eigenvalue inequality quoted in the proof of **Lemma 4.2** is follows readily that one can always assume that all the m_i in some vector space basis are positive definite. ❑

Given the set $\{a_i\}$, it follows that $\mathscr{S}(\{a_i\}) = \mathscr{S}(\{m_i\})$ for some basis set of matrices $\{m_i\}$. In most of our statistical applications it will usually be assumed that the m × m identity matrix $\mathbf{I_m}$ is contained in the set $\{a_i\}$, and so without loss of generality $\mathbf{I_m}$ is among the basis elements $\{m_i\}$. In all our applications though, we can always assume that $\mathscr{S}$, hence also $\{m_i\}$, contains some identity element $e \in \mathscr{S}$: the spaces used will be assumed to contain at least one positive definite element, hence an orthogonal transformation of the original basis set yields a new set of matrices, one of which is exactly I_m.

Recall from **Lemma 3.1** that e is always both an associative and a special Jordan identity for $\mathscr{S}$. Also, recall that by **Theorem 4.4** that a special, formally real Jordan algebra always contains an identity element.

Now let $\mathscr{S} = \mathscr{S}(\{m_i\})$ be given. We proceed to construct three new algebraic objects from the vector space $\mathscr{S} \subseteq S_m$:

(I)   First, let $\mathscr{A} = \mathscr{A}(\mathscr{S})$ be the smallest associative algebra, in $[\mathfrak{R}]_m$, that contains $\mathscr{S}$. As a practical matter it is the space of all polynomials, in all degrees, in all elements of $\mathscr{S}$ (since $\mathscr{S}$ contains an identity element e).

(II)   Second, in the vector space $\mathscr{S}$, consider the subspace $\mathscr{B} = \mathscr{B}(\mathscr{S})$ given by:

Definition. Given $\mathscr{S} = \mathscr{S}\{m_i\}$ let
$$\mathscr{B} = \mathscr{B}(\mathscr{S}) = \{b \in \mathscr{S} \mid sbs \in \mathscr{S}, \text{ for all } s \in \mathscr{S}\}.$$

Before plunging into the algebraic study of $\mathscr{S} = \mathscr{B}(\mathscr{S})$, we first try and remove some of the arbitrariness and mystery in its definition, by recalling its original statistical motivation.

It was found in Malley [1986] that, given a mixed linear model for data y having

$$y = \Sigma \, Z_i b_i, \quad \text{for } \{Z_i\} \text{ known design matrices;}$$

with

{b_i} = the set of random effects vectors, each with mean zero and all
        components of each b_i having kurtosis zero;

and

$$V = \text{var}(y) = \sum \sigma_i^2 \, Z_i Z_i^{\,T} \quad \text{with } \{\sigma_i^2\} \text{ the unknown variance components,}$$

then:

the space of all quadratic, unbiased estimates of the form $y^T b y$, with

$$b \in \mathscr{B}(\mathscr{S}), \quad \mathscr{S} = \mathscr{S}(\{Z_i Z_i^{\,T}\}),$$

is exactly the space of all optimal unbiased quadratic estimates for all estimable functions of the variance components $\{\sigma_i^2\}$, where *optimal* means: of minimum variance in the class of quadratic unbiased, translation invariant estimates.

For data y allowed to have non-zero mean, $E(y) = X\alpha$, (α the fixed effects parameter) a slight modification of $\mathscr{S}$ and $\mathscr{B}$ serves to locate all the unbiased, quadratic, translation invariant estimates of any estimable function of the components, as a certain subspace of the Jordan algebra $\mathscr{B}$.

A further simple modification leads to the basic result: all the optimal unbiased estimates for data having arbitrary kurtosis and mean, are generated by elements of a subspace located within $\mathscr{B}$. Another equally straightforward adjustment precisely locates (again in $\mathscr{B}$) the class of all optimal unbiased estimates for data having arbitrary mean and kurtosis, and random effects that may have any arbitrary non-zero correlation structure among themselves.

Let's return to a detailed algebraic study of $\mathscr{B}$.

In the language introduced in Chapter 3 below, $\mathscr{B}(\mathscr{S})$ is shown to be the **maximal ($\mathscr{S}$, $\mathscr{S}$)–outer**, and also **maximal ($\mathscr{S}$, $\mathscr{S}$)–inner**, subspace of $\mathscr{S}$. Leaving the precise definitions and technical details aside for now, an important consequence of this latter property of $\mathscr{B}$ is quite useful and justifies a separate reference to it here:

5.1 Theorem.   Given $\mathscr{S}$, $\mathscr{B} = \mathscr{B}(\mathscr{S})$ is the maximal subspace of $\mathscr{S}$ such that:
    $bsb \in \mathscr{B}$, for all $s \in \mathscr{S}$, $b \in \mathscr{B}$.
Proof. See Chapter 3. ■

Also of utility is the following (c.f. Malley [1986]):

5.2 Theorem.   Given $\mathscr{S}$, it follows that $\mathscr{B} = \mathscr{B}(\mathscr{S})$ is a finite dimensional, formally real, special Jordan algebra.
Proof. Using earlier observations about subspaces of S_m it only remains to show that $\mathscr{B}$ is Jordan, and this follows from **Lemma 4.3**. A more direct verification, though, makes for good practice with Jordan techniques.

Thus it is required to show that $g, h \in \mathscr{B}$ implies

$$g \cdot h = \tfrac{1}{2}(gh + hg) \in \mathscr{B}.$$

From the definition of B, $(c + d) \in \mathscr{B} \subseteq \mathscr{S}$, for all $c, d \in \mathscr{S}$, so that
    $(c + d)h(c + d) \in \mathscr{B}$,
and this holds if and only if
    $chd + dhc \in \mathscr{S}$, for all $c, d \in \mathscr{S}$, and $h \in \mathscr{B}$.
Since an identity $e \in \mathscr{S}$ is assumed,
    $age + ega = ag + ga \in \mathscr{S}$, for all $a \in \mathscr{S}$.
Thus
    $(ag + ga)h(a) + (a)h(ag + ga) \in \mathscr{S}$,
and this occurs if and only if
    $agha + gaha + ahag + ahga \in \mathscr{S}$,
so that
    $a(gh + hg)a + [(aha)g + g(aha)] \in \mathscr{S}$.

Finally, since $aha \in \mathscr{S}$, the second term above is in $\mathscr{S}$, which implies that
   $a[gh + gh]a = 2a(g \cdot h)a \in \mathscr{S}$,
for all $a \in \mathscr{S}$, and all $g, h \in \mathscr{B}$. The proof that $\mathscr{B}$ is Jordan is complete. ∎

Aside from the Jordan algebra $\mathscr{B} = \mathscr{B}(\mathscr{S})$, consider now a slightly weaker construction, one that also always leads to a Jordan algebra $\subseteq \mathscr{S}$:

Definition. Let $\mathscr{C}$ be a subspace of $\mathscr{S}$. $\mathscr{C}$ is said to be **j-closed** (Jordan closed), with respect to $\mathscr{S}$ if: $scs \in \mathscr{S}$, for all $c \in \mathscr{C}$, $s \in \mathscr{S}$.

According to **Theorem 5.1**, $\mathscr{B}$ is the **maximal j-closed subspace** of $\mathscr{S}$. Chapter 3 contains a more detailed look at subspaces closed with respect to several other product conditions.

(III)   The third construction is $\mathscr{J} = \mathscr{J}(\mathscr{S})$ = the smallest Jordan algebra in $\mathbf{S_m}$ that contains $\mathscr{S}$; we'll call it the **Jordan closure** of $\mathscr{S}$. Chapter 3 presents four algorithms for getting $\mathscr{J}$ after at most a finite number of steps, and involving no more than matrix multiplication and checks for linear independence. It is important to note that $\mathscr{J}$ is unique, and also important to observe that any subspace $\mathscr{E}$ in $\mathscr{J}$ always has a trace-product orthogonal complement in $\mathscr{J}$: there always exists a subspace $\mathscr{E}^{\perp} \subseteq \mathscr{J}$, such that:
   $\mathrm{trace}(\mathscr{E} \, \mathscr{E}^{\perp}) = \mathrm{tr}(\mathscr{E} \, \mathscr{E}^{\perp}) = 0$, and $\mathscr{J} = \mathscr{E} \oplus \mathscr{E}^{\perp}$.

Given the three constructions above **(I)**, **(II)**, and **(III)**, we have at hand the chain of subspaces:

$$\mathscr{B}(\mathscr{S}) \subseteq \mathscr{S} \subseteq \mathscr{J}(\mathscr{S}) \subseteq \mathscr{A}(\mathscr{S}) \cap \mathbf{S_m} \subseteq \mathscr{A}(\mathscr{S}),$$

where $\mathscr{A}(\mathscr{S})$ is the smallest associative algebra in $\mathbf{S_m}$ that contains $\mathscr{S}$.

Remark.   The precise nature of the inclusion

$$\mathscr{J}(\mathscr{S}) \subseteq \mathscr{A}(\mathscr{S}) \cap \mathbf{S_m}$$

is taken up in Chapter 3. In general the inclusion is strict, and this is important. It means that Jordan algebras generated by sets of symmetric matrices always in practice have strictly lower dimension than the corresponding associative algebras generated by the same set, except in the trivial case where the set of matrices all commute pair-wise. Simple examples show that $\dim \mathscr{J}(\mathscr{S})$ is usually much smaller than $\dim \mathscr{A}(\mathscr{S})$, so any statistical calculations using $\mathscr{J}$ will be usually be noticeably more tractable then one using $\mathscr{A}$. □

The next result can be of some help in explicitly calculating the Jordan closure $\mathscr{J}(\mathscr{S})$ and its decomposition into simple Jordan ideals, as many spaces $\mathscr{S}$ arising in practical statistical problems (e.g. mixed linear models having two or three variance components) will have $\dim \mathscr{S} \leq 3$.

5.3 Theorem.   Let $\{m_i\}$ be a basis set for $\mathscr{S}$, with $1 \leq i \leq k$, and where we assume $m_k = I$. Then:

(i) For $k = 1$, $\mathscr{S} = \mathscr{J}(\mathscr{S}) = \mathfrak{R}$;
(ii) For $k = 2$, $\mathscr{J}(\mathscr{S})$ is the commutative and associative algebra

$$\mathscr{A} = \mathscr{A}(\mathscr{S}) = \oplus \mathfrak{R} \; ,$$

where the direct sum has n terms, each equal to $\mathfrak{R}$, and where n is the degree of the minimum polynomial for m_1 . The ith component is the one dimensional eigenspace

$$\mathscr{S}_i = \mathscr{S}(\{E_i\}) = \mathfrak{R} \; ,$$

for E_i the idempotent projection associated with the ith eigenvalue of m_1 ;

(iii)   For a, b symmetric idempotents and let $\mathscr{S} = \mathscr{H}(a, b)$. Then $\mathscr{J}(\mathscr{S})$ is formed as follows. Let $t_1 = aba$, $t_2 = bab$, and let λ_i, $1 \leq i \leq n$, be the distinct nonzero eigenvalues of t_1 . Write:

$$f_1 = (t_1 - \lambda_1) \cdots (t_1 - \lambda_n) / \left[(-1)^n \lambda_1 \cdots \lambda_n \right]$$

$$f_2 = (t_2 - \lambda_1) \cdots (t_2 - \lambda_n) / \left[(-1)^n \lambda_1 \cdots \lambda_n \right]$$

$$g_1 = af_1 + f_1 a, \qquad g_2 = bf_2 + f_2 b$$

$$\xi_1^{(\alpha)} = t_1 [\prod_{j \neq \alpha} (t_1 - \lambda_j)] / [\lambda_\alpha \prod_{j \neq \alpha} (\lambda_\alpha - \lambda_j)],$$

$$\xi_2^{(\alpha)} = t_2 [\prod_{j \neq \alpha} (t_2 - \lambda_j)] / [\lambda_\alpha \prod_{j \neq \alpha} (\lambda_\alpha - \lambda_j)]$$

$$\psi_\alpha = \xi_1^{(\alpha)} \qquad \text{if } \lambda_\alpha = 1,$$

$$\psi_\alpha = \left(\xi_1^{(\alpha)} - \xi_2^{(\alpha)} \right)^2 / (1 - \lambda_\alpha) \qquad \text{if } \lambda_\alpha \neq 1.$$

Then the direct sum decomposition of $\mathscr{J} = \mathscr{J}(\mathscr{S}(a, b))$ into simple ideals is:

$$\mathscr{J}(\mathscr{S}) = \Re g_1 \oplus \Re g_2 \oplus (\oplus \Psi_\alpha),$$

for Ψ_α the ideal in $\mathscr{J}(\mathscr{S})$ generated by Ψ_α. One or both of $\Re g_1$, $\Re g_2$ may be zero. For $\lambda_\alpha = 1$, Ψ_α is the one dimensional algebra

$$\Re\left(\xi_1^{(\alpha)}\right).$$

For $\lambda_\alpha \neq 1$, Ψ_α is isomorphic to $\mathbf{S_2}$ and has a basis consisting of the the matrices $v_1^{(\alpha)}$, $v_2^{(\alpha)}$, $v_3^{(\alpha)}$:

$$v_1^{(\alpha)} = \left(1 - \lambda_\alpha\right)^{-1}\left[\xi_1^{(\alpha)} - \xi_2^{(\alpha)}\xi_1^{(\alpha)}\right]$$

$$v_2^{(\alpha)} = \left(1 - \lambda_\alpha\right)^{-1}\left[\xi_1^{(\alpha)}\xi_2^{(\alpha)} + \xi_2^{(\alpha)}\xi_1^{(\alpha)} - \lambda_\alpha\left(\xi_1^{(\alpha)} + \xi_2^{(\alpha)}\right)\right]$$

$$v_3^{(\alpha)} = \left(1 - \lambda_\alpha\right)^{-1}\left[\xi_2^{(\alpha)} - \xi_1^{(\alpha)}\xi_2^{(\alpha)}\right].$$

Proof. It is straightforward to see that only Case **(iii)** requires proof here. From Chapter 3 we will find that

$$\mathscr{S} = \mathscr{S}\{(m_i)\} \subseteq \mathscr{A}(\mathscr{S}) \cap \mathbf{S_m}$$

is the Jordan algebra generated by $\{m_i\}$ and by all the "tetrad" products in the m_i, $(m_i \neq \mathbf{I})$ of the form

$$m_{i_1} m_{i_2} m_{i_3} m_{i_4} + m_{i_4} m_{i_3} m_{i_2} m_{i_1} \qquad \text{for } i_1 < i_2 < i_3 < i_4.$$

Hence for

$$\mathscr{S} = \mathscr{S}(m_1, m_2, m_3, \mathbf{I})$$

we find

$$\mathscr{J}(\mathscr{S}) = \mathscr{A}(\mathscr{S}) \cap \mathbf{S_m},$$

since there are no tetrads of the required type (as described above) that require adjoining.

Thus, to calculate $\mathscr{J}(\mathscr{S})$ first calculate $\mathscr{A}(\mathscr{S})$, and then find in it all its symmetric elements. That is, first find $\mathscr{A}(\mathscr{S})$ and then form

$$\mathcal{J}(\mathcal{S}) = \{a + a^T \mid a \in \mathcal{A}(\mathcal{S})\} \, .$$

Still more explicitly we can modify Theorem 6 of Mann [1960; p. 5], where it is shown how to calculate $\mathcal{A}(\mathcal{S})$ and its direct sum decomposition into associative simple ideals, **in the special case that m_1 and m_2 are idempotents,**

$$m_1{}^2 = m_1, \quad m_2{}^2 = m_2 \, .$$

Using the notation of Mann [1960] let $p_1 p_1{}^T = m_1$, $p_2 p_2{}^T = m_2$. It also helps, in reading Mann [1960], to recall that the distinct nonzero eigenvalues of uu^T, for u any s × t matrix, are exactly those of $u^T u$ (c.f. Anderson [1984; p. 531]). Hence for symmetric idempotents a and b the eigenvalues of

$$aba = ab^2 a = ab(ab)^T$$

are exactly those of

$$(ab)^T ab = ba^2 b = bab \, .$$

At this point the argument of Mann [1960], used for calculating

$$\mathcal{A}(\mathcal{S}(m_1, m_2))$$

goes through with only obvious modifications. This mainly involves straight-forward checking that the proper quantities are indeed orthogonal idempotents as required. The proof of Case (iii), and hence the Theorem, is now complete. ∎

Remark. $\mathcal{J}(\mathcal{S}(a, b))$ in the Theorem above is at most five dimensional. □

Remark. The process of getting idempotents g given $a \in \mathbf{S}_m$, with the property that ag = ga, is often useful in itself. An important statistical example of this process occurs later in Section 2.6, where we link up Case (iii) above with the general problem of independence for random forms $y^T ay$, $y^T by$ for arbitrary a, b $\in \mathcal{J}(\mathcal{S})$. This can be done using Jacobson [1968; p. 149, Lemma 1], or simply as follows. Given a symmetric s write e = ss⁺. Then:

$$e^2 = (ss^+)(ss^+) = (ss^+ s)s^+ = ss^+ = e \, ,$$

Once again the idempotent e gets expressed as a polynomial in s, with

$$e \in \mathcal{J}(\mathcal{S}(s)) \, . \quad □$$

To conclude this Section, it is of independent interest to pin down the relationship between subspaces, and associative or Jordan ideals, particularly with respect to our $\mathscr{S}, \mathscr{J}$ and $\mathscr{A}$. McCrimmon [1969] has studied these questions for not necessarily special, or formally real Jordan algebras, while Herstein [1969] has looked at them from a still wider, ring-theoretic perspective. Malley [1986] gives several results that are more specific to our context, including the following; it also appears as Lemma 1, p. 10, of Jacobson [1968], while the proof given in Chapter 3 is in our opinion more direct and elementary:

5.4 Theorem. Suppose $\mathscr{C}$ is a subspace of $\mathscr{J} = \mathscr{J}(\mathscr{S})$. Then $\mathscr{C}$ is a Jordan ideal of $\mathscr{J}$ if and only if $\mathscr{C}$ contains all the triple products scs, for all $s \in \mathscr{S}$, and all $c \in \mathscr{C}$.

2.6 The Algebraic Study of Random Quadratic Forms.

We now present the algebraic results that follow from the constructions and facts given above and that have application to the study of random quadratic forms.

Let's begin by saying that for any space $\mathscr{S}$ having a direct sum decomposition $\mathscr{S} = \oplus \mathscr{S}_i$, the **support** of $a \in \mathscr{S}$, $a = \oplus a_i$, $a_i \in \mathscr{S}_i$, is the set of indicies i such that $a_i \neq 0$. For $a, b \in \mathscr{S}$, then, a and b are said to have **disjoint support** if and only if $a_i \neq 0$ implies $b_i = 0$, and conversely, $b_i \neq 0$ implies $a_i = 0$. Applying **Theorem 4.4** we see that $\mathscr{B}(\mathscr{S}) = \oplus \mathscr{B}_i$ for a set $\{\mathscr{B}_i\}$ of simple Jordan ideals of $\mathscr{B}(\mathscr{S})$. Then:

6.1 Theorem. Suppose $\mathscr{S} \subseteq \mathbf{S}_m$, and
$$\mathscr{B} = \mathscr{B}(\mathscr{S}) = \{b \in \mathscr{S} \mid sbs \in \mathscr{S}, \text{ all } s \in \mathscr{S}\},$$
with $\mathscr{S}$ having an identity element, then for any a and $b \in \mathscr{B}$, the following are equivalent:

    (i)   sasbs = 0   for all $s \in \mathscr{S}$;

    (ii) $a \mathscr{S} b = 0$;

    (iii) $a \mathscr{B} b = 0$;

    (iv) a and b have disjoint support in $\mathscr{B}$.

The proof appears in Chapter 3. For now we note that the equivalence of **(i)** and **(ii)** is a non-Jordan, purely manipulative fact, and so can be stated and proven in a much wider context:

6.2 Theorem. Let G be any set with a multiplicative identity element e in some associative algebra $\mathscr{A}$. Suppose there are two fixed, real constants $\alpha, \beta > 0$, $\alpha \neq \beta$, such that for all $g \in G$, and some fixed $\alpha, \beta \in \mathscr{A}$:

$$(\alpha + 1)^{-1}[\alpha g + e] \qquad \text{and} \qquad (\beta + 1)^{-1}[\beta g + e] \in G .$$

Then gagbg = 0 for all $g \in G$ if and only if aGb = 0.

Proof. Letting $g = e$ in gagbg = 0 gives ab = 0. Since $t = \delta g + e$ is such that

$(\alpha + 1)^{-1} t$ or $(\beta + 1)^{-1} t \in G,$

for $\delta = \alpha$ or β respectively, we get that

$(a + 1)^3 \, tatbt = 0$ or $(\beta + 1)^3 \, tatbt = 0.$

In either case $tatbt = 0$, so that:

$0 = (\delta g + e) \, a \, (\delta g + e) \, b \, (\delta g + e)$

$= \delta \, [(\delta g + e) \, agb \, (\delta g + e)]$ (since $ab = 0$)

$= \delta \, [\delta \, (gagb + agbg) + agb].$

But $\delta \neq 0$ gives

$agb = -\delta(gagb + agbg)$ for $\delta = \alpha$ or β,   with $\alpha \neq \beta$,

so that only

$gagb + agbg = agb = 0$

is possible, and $aGb = 0$. Since we always have $aGb = 0$ implies $gagbg = 0$, the proof is complete. ∎

Using methods akin to those of **Theorems 6.1** and **6.2** we also get:

6.3 Theorem. Given $\mathscr{S} \subseteq \mathbf{S}_m$, $\mathscr{J} = \mathscr{J}(\mathscr{S})$ the Jordan closure of $\mathscr{S}$, with $\mathscr{S}$ having an identity element, then for any elements a, b in $\mathscr{J}$ the following are equivalent:
- **(i)** $sasbs = 0$ for all $s \in \mathscr{J}$;
- **(ii)** $a\mathscr{S}b = 0$;
- **(iii)** a and b have disjoint support in $\mathscr{J} = \oplus \mathscr{J}_i$.

Remark. Condition (i) refers to all $s \in \mathscr{J}$ and not merely all $s \in \mathscr{S}$. We would expect that this could be a significant restriction, but it is not.

First, we show below that by merely making the additional assumption that $I \in \mathscr{S}$ it follows that both the conditions can be dropped and that $a\mathscr{S}b = 0$ implies $a\mathscr{J}b = 0$. From this the other two statements, (ii) and (iii), are immediate.

Second, for essentially all mixed linear model problems it can be assumed that the identity matrix $I \in \mathscr{S}$. This obtains since for these models it is common to assume that the collection of allowable covariance matrices has at least one positive definite element. Then by a suitable orthogonal rotation of the data one gets I as an allowable covariance matrix, and $I \in \mathscr{S}$. No loss of generality occurs, since for any orthogonal matrix u, it follows that any space $\mathscr{J}$ of symmetric matrices is Jordan if and only if $u^T \mathcal{J} u$ is Jordan, and that, for example,

$a \mathcal{S} b = 0$

if and only if

$(u^T au)(u^T \mathcal{S} u)(u^T bu) = a \mathcal{S} b = 0.$ □

Remark. The condition that a and $b \in \mathcal{J}$ may also seem restrictive. Again, for the mixed models usually considered, testing the independence pairs of sums of squares, and the χ^2–distributivity of a single random quadratic form, will usually involve random quadratic forms in normal variables of the form $y^T ay$ with a an element of $\mathcal{J}$. □

We defer the proof of **Theorem 6.3** to Chapter 3, but note that if the decomposition of $\mathcal{S}$ is obtained from one of $\mathcal{J}$,

$$\mathcal{S} \subseteq \mathcal{J} = \oplus \mathcal{J}_i, \quad \mathcal{S} = \oplus \mathcal{S}_i, \quad \mathcal{S}_i \subseteq \mathcal{J}_i,$$

then a and b having disjoint support in $\mathcal{S}$ immediately implies

$a \mathcal{S} b = a \mathcal{J} b = 0.$

Also, if $ab = 0$, then the converse of this is true and relatively easy to prove. Note that the condition $ab = 0$ is immediate if $I \in \mathcal{S}$. Recall now the following standard result:

6.4 Lemma. For matrices a, b, c and $d \in [\Re]_m$, with b and d positive semide-finite:

$$[\operatorname{tr}(abcd)]^2 \leq [\operatorname{tr}(abad)][\operatorname{tr}(bcbd)].$$

Proof. Consider the bilinear functional
   $(a, c) = \operatorname{tr}[abcd].$
As b and d are positive semidefinite the inner product is also positive semi-definite.
   Verifying this requires showing that $(a, a) \geq 0$ for all a. But the fact that b and d are positive semidefinite implies there exist square matrices g and h such that

$b = gg^T,$ and $d = hh^T.$

Write $u = [u_{ij}] = h^T ag.$ Then:

$(a, a) = \text{tr}[abad]$

$\qquad = \text{tr}[agg^T ahh^T]$

$\qquad = \text{tr}[(h^T ag)(h^T ag)^T]$

$\qquad = \text{tr}[uu^T] = \Sigma(u_{ij})^2 \geq 0,$

as required.

Finally, applying the Cauchy-Schwarz inequality (see for example Eaton [1983; p. 26]), completes the proof. ∎

6.5 Theorem. Suppose a and $b \in \mathbf{S}_m$ and that $ab = 0$. Then:

$$a\mathscr{S}b = 0 \qquad \text{if and only if} \qquad a\mathscr{J}(\mathscr{S})b = 0.$$

Proof. Referring to results in Chapter 3 on the construction of $\mathscr{J}(\mathscr{S})$ from $\mathscr{S}$ it can be checked that it is enough to show:

$$asb = 0 \quad \text{for all } s \in \mathscr{S} \text{ implies} \quad as^2 b = 0.$$

Now, by **Lemma 4.3**, $as^2 b = 0$ if and only if

$$\text{tr}[(as^2 b)(bs^2 a)] = \text{tr}[(as^2 b)(as^2 b)^T] = 0,$$

and this is equivalent to showing

$$\text{tr}[a^2 s^2 b^2 s^2] = \text{tr}[(s^2 a^2)s(s)b^2] = 0.$$

Using the eigenvalue argument appearing in the proof of **Lemma 4.2**, for any given fixed $s \in \mathscr{S}$, $s \neq 0$, there exists a ζ such that $t = \mathbf{I} - \zeta s$ is positive definite; b^2 is always p.s.d. Hence **Lemma 6.4** shows that

$$(c, d) = \text{tr}[(c)t(d)b^2]$$

is a positive semidefinite inner product, and that

$$\{\text{tr}[stdb^2]\}^2 \leq (\text{tr}[ctcb^2])(\text{tr}[dtdb^2]).$$

But

$$\text{tr}[ctdb^2] = \text{tr}[c(\mathbf{I} - \zeta s)db^2]$$

$$\qquad = \text{tr}[cdb^2] - \zeta\text{tr}[csdb^2],$$

and putting $c = s^2a$, $d = s$ yields:

$$tr[s^2a^2sb^2] - \zeta tr[s^2a^2s^2b^2] = -\zeta tr[s^2a^2s^2b^2],$$

since

$$0 = a(asb)b = a^2sb^2.$$

Thus

$$\{-\zeta tr[s^2a^2s^2b^2]\}^2 \leq (tr[s^2a^2(\mathbf{I} - \zeta s)a^2s^2b^2])(tr[s(\mathbf{I} - \zeta s)sb^2]).$$

However

$$tr[s^2a^2(\mathbf{I} - \zeta s)s^2a^2b^2] = 0$$

as $a\mathscr{S}b = 0$ with $\mathbf{I} \in \mathscr{S}$ implies

$$a^2b^2 = a(a(\mathbf{I})b)b = 0.$$

Thus

$$tr[a^2s^2b^2s^2] = tr[(as^2b)(bs^2a)^T] = 0$$

which by **Lemma 4.3** implies $as^2b = 0$, and the proof is complete. ∎

6.6 Theorem. Given $\mathscr{S} \subseteq \mathbf{S}_m$, with $\mathbf{I} \in \mathscr{S}$, and $\mathscr{J} = \mathscr{J}(\mathscr{S})$ the Jordan closure of $\mathscr{S}$, and elements a and b in $\mathscr{J}$, then the following are equivalent:
(i) $sasbs = 0$ for all $s \in \mathscr{J}$;
(ii) $a\mathscr{S}b = 0$;
(iii) $a\mathscr{J}b = 0$;
(iv) a and b have disjoint support in $\mathscr{J}$.

Proof. Follows from **Theorems 6.3** and **6.5**. ∎

Finally, another decomposition result, easily verified in view of the results above, is:

6.7 Theorem. Given $\mathscr{S} \subseteq \mathbf{S}_m$, $\mathscr{J} = \mathscr{J}(\mathscr{S}) = \oplus\mathscr{J}_i$, for ideals $\mathscr{J}_i \subseteq \mathscr{J}$, then:

$$sasas = sas \quad \text{for all } s \in \mathscr{S} \quad \text{if and only if} \quad sa_isa_is = sa_is$$

for all i and all $s \in \mathscr{S}_i$, with $a = \oplus a_i$, $a_i \in \mathscr{J}_i$.

The reader can now probably see the outlines of results for the independence and χ^2–distributivity of random quadratic forms that use direct sum ideal decompositions of Jordan algebras. We fill in the details in the next section, with the main parts of our algebraic journey now mostly behind us.

2.7 The Statistical Study of Random Quadratic Forms.

We now describe what may be viewed as maximal extensions of the classical result of Cochran [1934], and the work of Rao and Mitra [1971] on the independence and χ^2–distributivity of quadratic forms in normal random variables.

In order to apply the algebraic results derived above we first quote two standard results on random quadratic forms. Assume that the random vector y is normally distributed, $y \sim \mathcal{N}(\mu, \mathcal{V})$, with mean μ and covariance matrix $\mathcal{V}$, where $\mathcal{V}$ is only required to be positive semidefinite. Then for real symmetric matrices a and b:

7.1 Theorem (Rao and Mitra [1971; Theorem 9.4.1]). Let $y \sim \mathcal{N}(\mu, \mathcal{V})$. Then the quadratic forms $y^T a y$ and $y^T b y$ are statistically independent if and only if:
$$\mathcal{V}a\mathcal{V}b\mathcal{V} = 0, \quad \mathcal{V}a\mathcal{V}b\mu = \mathcal{V}b\mathcal{V}a\mu = 0, \quad \mu^T a\mathcal{V}b\mu = 0.$$

Also well-known is:

7.2 Theorem (Rao and Mitra [1971; Theorem 9.2.1]). For $y \sim \mathcal{N}(\mu, \mathcal{V})$, a necessary and sufficient condition that the random quadratic form $y^T a y$ be distributed as a central χ^2, on $k = \mathrm{tr}(a\mathcal{V})$ degrees of freedom, is that:
$$\mathcal{V}a\mathcal{V}a\mathcal{V} = \mathcal{V}a\mathcal{V}.$$

Assume it what follows, unless otherwise stated, that

$$\mathscr{S} = \mathscr{S}(\mathcal{V}) = \text{the space in } \mathbf{S_m} \text{ spanned by the set of all allowable } \mathcal{V}.$$

Now, it has been shown that given any space $\mathscr{S} \subseteq \mathbf{S_m}$, then $\mathscr{J} = \mathscr{J}(\mathscr{S})$, has a highly structured interior. It orthogonally decomposes into an ideal direct sum of simple Jordan ideals, and these in turn come in just four flavors, those cases appearing in **Theorem 4.4**. Completely parallel facts obtain for any j–closed subspace $\mathscr{S}_0 \subseteq \mathscr{S}$, in particular for $\mathscr{B}(\mathscr{S}) \subseteq \mathscr{S}$. The translation of this rich interior of $\mathscr{B}$ and of $\mathscr{J}$, into statistically useful statements begins with the following observations.

The two theorems above, **7.1** and **7.2**, are stated in terms of a given, known covariance matrix $\mathcal{V}$. Our attention in what follows will be focused on a **set** of allowable covariance matrices. These matrices may be merely positive semidefinite and not required to be all positive definite. In any event, the linear space spanned by the set of all allowable covariance matrices will contain matrices that are neither positive semidefinite nor allowable covariance matrices.

On the other hand, one of our goals is to find necessary and sufficient conditions for statistical independence (for pairs of quadratic forms) and for χ^2–distributivity (for a single quadratic form). Such conditions are most easily found and stated in terms of conditions involving the linear space spanned by the covariance matrices. But, and this is the point we wish to make here, this does not materially limit the classes of sets of allowable covariance matrices for which the results of the theorems hold. More precisely, after presenting several results stated in terms of the linear space spanned by the set, and the Jordan algebra generated by that space, we go on in the next section to show that if the conditions hold for a collection of allowable covariance matrices that is a covex set, then they must hold as well for the linear space spanned by the set. And convexity of a set of allowable covariance matrices is often a scientifically reasonable condition. It is also a mathematically reasonable one, in as much as a convex combination of elements of a set of positive definitive matrices is again positive definite; a parallel statement holds for positive semidefinite matrices.

Finally, since the linear space spanned by a convex set of symmetric matrices (the linear hull of the convex set) must have finite dimension, it follows by inspection that **the conditions of the results stated below obtain if and only if they do so for a basis of the linear hull**. Note particularly that the basis set is **not** required to be composed of only positive definite elements, or even consist of allowable covariance matrices. Hence checking the conditions of the theorems reduces to checking them on **any** set of basis matrices for the linear space generated by the convex set in question.

The results to be presented, therefore, utilize both scientifically reasonable and numerically tractable conditions to obtain the necessary and sufficient conclusions obtained. We formalize the preceeding discussion in **Theorem 8.2** below.

Let's begin with:

7.3 Theorem. The simple Jordan ideal decomposition of $\mathscr{B} = \mathscr{B}(\mathscr{S}) = \oplus \mathscr{B}_i$, constitutes a **maximal independence decomposition** of $\mathscr{S}$:

   (i) b_i, $b_i^* \in \mathscr{B}_i$, both $\neq 0$, implies $y^T(b_i)y$ and $y^T(b_i^*)y$ are statistically independent for all $\mathscr{V} \in \mathscr{S}$;

   (ii) $b_i \in \mathscr{B}_i$, $b_j \in \mathscr{B}_j$, $i \neq j$, implies $y^T(b_i)y$ and $y^T(b_j)y$ are statistically independent for all $\mathscr{V} \in \mathscr{S}$.

Proof. The is a straightforward application of **Theorems 6.4** and **7.1**. ∎

Remark. Recall that $ab = 0$ is assured if $\mathbf{I} \subseteq \mathscr{S}$. This condition on the linear space $\mathscr{S} = \mathscr{S}(\mathscr{V})$ spanned by all the allowable covariance matrices $\mathscr{V}$ is essentially always valid, since as observed in a previous remark, the matrix identity $\mathbf{I}$ can generally be assumed a member of the set $\mathscr{V}$. □

It may well be that a given $\mathscr{S}$ has $\mathscr{B}(\mathscr{S}) = 0$, a case of this occuring in the one-way, unbalanced, random effects model: see Malley [1986; p. 68]. Hence we give a version of Theorem 7.3 that makes no reference to $\mathscr{B}(\mathscr{S})$:

7.4 Theorem. Suppose $a, b \in \mathscr{J}(\mathscr{S})$. Then the random forms $y^{\mathrm{T}}ay$, $y^{\mathrm{T}}by$ are statistically independent if and only if they have disjoint support in
$$\mathscr{J} = \mathscr{J}(\mathscr{S}) = \oplus \mathscr{J}_i .$$

Proof. If a, b have disjoint support in $\mathscr{J}$, then it follows that they have disjoint support in the decomposition for $\mathscr{S}$ derived from that of $\mathscr{J} = \oplus \mathscr{J}_i$, that is
$$\mathscr{S} = \oplus \mathscr{S}_i , \quad \mathscr{S}_i \subseteq \mathscr{J}_i .$$

Hence for all $s \in \mathscr{S}$ we get $asb = 0$, by **Theorem 6.3.** For all covariance matrices $\mathscr{V} \in \mathscr{S}$ it then is true that $a \mathscr{V} b = 0$, and also $\mathscr{V} a \mathscr{V} b \mathscr{V} = 0$, so independence follows.

The converse is clear since then $a \mathscr{S} b = a \mathscr{J}(\mathscr{S}) b = 0$, by **Theorem 6.6,** and this completes the proof. ∎

Remark. Theorems **7.3** and **7.4** are stated as decomposition results in order to highlight the way they generalize the classical result of Cochran. Recall though that in each a more practical equivalent formulation uses **Theorems 6.3** and **6.5.** Then a necessary and sufficient condition for independence assumes the form: $y^{\mathrm{T}}ay$ and $y^{\mathrm{T}}by$ are statistically independent if and only if $a(m_i)b = 0$ for all m_i that form a basis for the space $\mathscr{S}$ generated by the set of all allowable covariance matrices. ◻

We can replace "independence" by merely "uncorrelated" in all the above, and also remove the normality condition: to accomplish this assume the data y has mean and kurtosis both equal to zero, relinquish the complete (if and only if) equivalances above, and use:

7.5 Theorem. Suppose $E(y) = 0$, and y has zero kurtosis. Then:
$$\mathrm{cov}(y^{\mathrm{T}}ay, y^{\mathrm{T}}by) = 2\mathrm{tr}(a \mathscr{V} b \mathscr{V}).$$

This is a standard result; see for example Malley [1986; p. 23]. Using it yields:

7.6 Theorem. Suppose the data y (not necessarily normal) has $E(y) = 0$, and kurtosis equal to zero. Then for $a, b \in \mathscr{J}$ the random forms $y^{\mathrm{T}}ay$, and $y^{\mathrm{T}}by$ are uncorrelated whenever a and b have disjoint support in
$$\mathscr{J} = \mathscr{J}(\mathscr{S}) = \oplus \mathscr{J}_i .$$

Proof. Disjoint support in $\mathscr{J}$ implies $asb = 0$ for all $s \in \mathscr{S}$, hence $a \mathscr{V} b = 0$ for all $\mathscr{V} \in \mathscr{S}$, and $\mathrm{tr}(a \mathscr{V} b \mathscr{V}) = 0$ as needed. ∎

Slightly different proofs of many of the results above, and other new facts as well, can be seen to follow from the orthogonality of the decomposition of $\mathscr{J}$. Thus partitioning the data vector y conformably with that of $\mathscr{J}$, via u, and the u_i, leads to $y = (y_1^T, \ldots, y_k^T)^T$ and:

$$uyu^T = z = [z_1^T, \ldots, z_k^T]^T, \quad \text{with} \quad u_i y_i u_i^T = z_i .$$

Parallel to our earlier definition of support let's agree to say that two Borel functions

$$u = u(y_1, \ldots, y_k), \quad v = v(y_1, \ldots, y_k),$$

have disjoint support in $\mathscr{J}$ if the indicies of the y_i appearing in u and v are disjoint. Then:

7.7 Theorem. Suppose $y \sim \mathscr{N}(\mu, \mathscr{V})$, $\mathscr{S} = \mathscr{S}(\mathscr{V})$, $\mathscr{J} = \mathscr{J}(\mathscr{S})$. Then any Borel functions w and z of the y_i are statistically independent if and only if they have disjoint support in $\mathscr{J}$.

Proof. $\mathscr{J} = \mathscr{J}(\mathscr{S}) = \oplus \mathscr{J}_i$ implies $u \mathscr{V} u^T = \oplus \mathscr{V}_i$, for an orthogonal u. Write $t = uyu^T$, and partition t conformably with the decompositon of $\mathscr{J}$. Then

$$\text{var}(uyu^T) = \text{var}(t) = u \mathscr{V} u^T = \text{diag}[\mathscr{V}_i, \ldots, \mathscr{V}_k].$$

Independence for the Borel functions w, z then follows since

$$\text{cov}(t_i, t_j) = 0, \quad \text{for } i \neq j,$$

and since each t_i and t_j is an invertible Borel function of y_i and y_j respectively. This completes the proof. ■

The ultimate merit of **Theorem 7.7** hinges, of course, on our capacity to calculate $\mathscr{J}(\mathscr{S})$, and the orthogonal decomposition of $\mathscr{J}(\mathscr{S})$. We expect this to be difficult for most $\mathscr{S}$. We hence proceed instead to give an alternative, interesting necessary and sufficient condition for the statistical independence of two quadratic forms, by a method using results given in **Theorem 2.4**, and the explicit construction in **Theorem 4.3(iii)**.

Write a^+ and b^+ for the Moore–Penrose inverses of a and b respectively, and begin with the following observation:

7.8 Lemma. Let $\mathscr{S}$ be any subspace of $\mathbf{S}_m$ and a, b any elements of $\mathbf{S}_m$. Then

$$a \mathscr{S} b = 0 \quad \text{if and only if} \quad (aa^+)\mathscr{S}(bb^+) = 0.$$

Proof. We know that $aa^+ = a^+a$, $bb^+ = b^+b$, and that aa^+ is a polynomial in a, and bb^+ is a polynomial in b. In each case we see by inspection that these polynomials have zero constant term.

Now suppose first that $(aa^+)\mathscr{S}(bb^+) = 0$. Then

$$
\begin{aligned}
0 &= a[(aa^+)\mathscr{S}(bb^+)]b \\
&= (aaa^+)\mathscr{S}(bb^+b) \\
&= (aa^+a)\mathscr{S}(bb^+b) \\
&= a\mathscr{S}b.
\end{aligned}
$$

For the other direction assume that $a\mathscr{S}b = 0$. Then for any polynomials $p(x)$, $q(x)$ having zero constant term, it follows that

$$p(a)\mathscr{S}q(b) = 0,$$

as can be verified by expanding the product. In particular then

$$(aa^+)\mathscr{S}(bb^+) = 0,$$

as required, and our proof is complete. ∎

Consider next the Jordan algebra $\mathscr{J}(\mathscr{S}_+)$ where

$$\mathscr{S}_+ = \mathscr{S}(aa^+, bb^+, \mathbf{I}).$$

In general

$$\mathscr{S} = \mathscr{S}(a, b, \mathbf{I})$$

will **not** be properly contained in $\mathscr{J}(\mathscr{S}_+)$, but both $\mathscr{S}$ and $\mathscr{J}(\mathscr{S}_+)$ will always be contained in $\mathscr{J}(\mathscr{S})$. On the other hand $\mathscr{J}(\mathscr{S}_+)$ will always have the orthogonal decomposition into simple Jordan ideals as given in **Theorem 4.6(iii)**. In particular for arbitrary elements of $\mathscr{J}(\mathscr{S}_+)$ we can readily calculate their support over the simple components $\mathscr{J}_i$ of $\mathscr{J}(\mathscr{S}_+)$. Indeed, aa^+ and bb^+ have disjoint support over $\mathscr{J}(\mathscr{S}_+)$ if and only if $\mathscr{J}(\mathscr{S}_+) = \mathscr{S}_+$, if and only if we have $aa^+bb^+ = 0$, if and only if $ab = 0$, since

$$
\begin{aligned}
0 &= a(aa^+)(bb^+)b \\
&= (aa^+a)(bb^+b) = ab.
\end{aligned}
$$

Turning now to the question of χ^2–distributivity for random quadratic forms, we find that in light of the results above:

7.9 Theorem. Suppose $y \sim \mathcal{N}(\mu, \mathcal{V})$ and that $a \in \mathcal{J}(\mathcal{S})$. Then the random form $y^T a y$ has a χ^2 distribution, with degrees of freedom equal to $\text{tr}(a\mathcal{V})$ if and only if every member of the decomposition of $y^T a y$ for $a \in \mathcal{J}$ has a χ^2 distribution, that is, if and only if every $y^T(a_i)y$, with $a = \oplus a_i$, $a_i \in \mathcal{J}_i$, has a χ^2 distribution, having degress of freedom equal to

$$\text{tr}(a\mathcal{V}) = \Sigma \, \text{tr}(a_i \mathcal{V}_i), \quad \mathcal{V}_i \in \mathcal{J}_i .$$

This represents our maximal statistical extension of the classical theorem of Cochran on the decomposition of χ^2–distributed variables.

Before proceeding to further applications of the material developed so far, we show how to generalize the facts obtained so far for linear spaces $\mathcal{S} = \mathcal{S}(\mathcal{V})$ to the situation in which the covariance matrices are constrained to lie in some convex, not necessarily linear space in $\mathbf{S}_m$. Convexity is often a reasonable, but still quite inclusive, request of a set of covariance matrices for the data, yet that the assumption of convexity leads to exactly the same collection of results as the assumption of linearity.

2.8 Covariance Matrices Restricted to a Convex Space.

Let space $\mathscr{C} \subseteq \mathbf{S}_m$ be convex: $\alpha c + \beta d \in \mathscr{C}$ for all real α, β such that

$$\alpha + \beta = 1, \quad \alpha, \beta \geq 0, \quad \text{and all } c, d \in \mathscr{C}.$$

The independence results of the last section are recovered through:

8.1 Lemma (The Linearizing Lemma). Suppose, for fixed $a, b \in \mathbf{S}_m$, that the equation
$$cacbc = 0$$
holds for all $c \in \mathscr{C} \subseteq \mathbf{S}_m$, where $\mathscr{C}$ is a convex space containing an identity element $e \in \mathscr{C}$ ($ec = ce = c$ for all $c \in \mathscr{C}$). Then the equation
$$sasbs = 0$$
holds for all $s \in \mathscr{S}$, where $\mathscr{S}$ is the linear span of $\mathscr{C}$.

Proof. First, putting $c = e$ in $cacbc = 0$ gives $ab = ba = 0$, while using $2d = c + e$ for any $c \in \mathscr{C}$, has $d \in \mathscr{C}$ and

$$0 = dadbd$$

$$= ab + abc + cacb + cacbc + cab + cabc + acb + acbc$$

$$= acb + acbc + cacb,$$

since $ab = 0$, so that

$$0 = (acb + acbc + cacb)a$$

$$= acbca.$$

Next given any $c \in \mathscr{C}$, and $acbca = 0$, we show $asbsa = 0$ for $s \in \mathscr{S}$, the linear span of $\mathscr{C}$, and this we do by verifying it for the particular $s = c + \gamma d$, for arbitrary real γ, and any $c, d \in \mathscr{C}$. By convexity

$$\tfrac{1}{2}(c + d) \in \mathscr{C},$$

so that

$$0 = a(c + d)b(c + d)a$$

$$= acbca + acbda + adbca + adbda$$

while

$$acbca = adbda = 0$$

by assumption on $c, d \in \mathscr{C}$. Hence

$$0 = \gamma(acbda + adbca)$$

$$= acb(\gamma d)a + a(\gamma d)bca.$$

Since $0 = adbda$ it follows that $\gamma^2(adbda) = 0$, and

$$0 = adbda + \gamma^2(adbda)$$

$$= acbca + a(\gamma d)b(\gamma d)a.$$

Adding the last two equations:

$$0 = adbda + a(\gamma d)b(\gamma d)a + acb(\gamma d)a + a(\gamma d)bca$$

$$= a(c + \gamma d)b(c + \gamma d)a.$$

The proof will be complete as soon as we can show that $asbsa = 0$ is equivalent to $sasbs = 0$, for all $s \in \mathscr{S}$. But **Theorem 5.2** applies here, with $\mathscr{G} = \mathscr{C}$, giving

$$a\mathscr{S}b = 0 \quad \text{for } \mathscr{S} \text{ the linear span of } \mathscr{C},$$

so also $sasbs = 0$ for all $s \in \mathscr{S}$. The proof is now finished. ∎

Hence, for example,

$\mathscr{V} a \mathscr{V} b \mathscr{V} = 0$ for all $\mathscr{V} \in \mathscr{C}$, with $\mathscr{C}$ convex,

implies

$\mathscr{V} a \mathscr{V} b \mathscr{V} = 0$ for all $\mathscr{V} \in \mathscr{S}$, with $\mathscr{S}$ the linear span of $\mathscr{C}$.

As the converse is clear, we have succeeded in "linearizing" the convex covariance set restriction, in so far as independence of random quadratic forms in normal variables is concerned. More formally:

8.2 Theorem. Assume $y \sim \mathscr{N}(\mu, \mathscr{V})$. Let the set $\mathscr{C}$ of all allowable covariance matrices $\mathscr{V}$ be convex, and write $\mathscr{S} = \mathscr{S}(\mathscr{C})$ be the linear space spanned by $\mathscr{C}$. Let $\{m_i\}$ $(1 \le i \le k)$ be a basis for $\mathscr{S}$. Then the random quadratic forms $y^T a y$ and $y^T b y$ are statistically independent if and only if $a(m_i)b = 0$ for all i.

Proof. Use **Theorems 6.6, 7.4** and **8.1.** ∎

2.9 Applications to the General Linear Mixed Model.

We turn now to showing how the results above simplify and unify previous work by James, Mann, Speed, Dawid and others, on the general linear model, and the analysis of structured data. A special feature of our analysis is that hypotheses testing of fixed effects in the model are explicitly linked to quadratic forms produced from idempotents that arise in an ideal-direct sum decomposition of the Jordan algebra uniquely generated by the fixed effects design matrix for the model.

In Mann [1960] it is shown how every linear hypothesis test arises from a decomposition of the identity element of a certain associative algebra derived from the linear model at hand. This algebra is that generated in the space of all real symmetric matrices by a submodel partitioning of the design matrix X.

That is, as Mann [1960] explains, let $\mathscr{U}$ be the associative algebra generated in real $m \times m$ matrices by the matrices $X_i X_i^T$ for:

$\mu = E(y) = X\beta$, $\beta = [\beta_1^T, \beta_2^T, \ldots, \beta_k^T]^T$,

$\mathscr{V} = \text{var}(y) = \sigma^2 I$,

with X partitioned as

$X = [X_1 \mid X_2 \mid \ldots \mid X_k]$.

Hence

$$E(y) = \Sigma\, X\beta = X_1\beta_1 + \ldots + X_k\beta_k.$$

Next, define the matrices

$$Z_i = [X_i \mid X_{i+1} \mid \ldots \mid X_k]$$

and define the idempotents (projections)

$$P_i = Z_i[Z_i^TZ_i]^-Z_i^T,$$

where for matrix A, a g–inverse of A is A⁻. With these P_i the identity matrix **I** is partitioned as

$$I = (I - P_1) + (P_1 - P_2) + \ldots + (P_{k\text{-}1} - P_k) + P_k,$$

and the total variance y^Ty is partitioned into separate sums of squares:

$$y^Ty = y^T(I)y$$

$$= y^T(I - P_1)y + y^T(P_1 - P_2)y + \ldots$$

$$\ldots + y^T(P_{k\text{-}1} - P_k)y + y^TP_ky.$$

Then by standard linear model theory (see e.g. Seber [1980]) the required orthogonal idempotents for hypotheses testing are selected from the list:

$$a_1 = I - P_1, \qquad a_2 = P_1 - P_2, \quad \ldots \quad, \qquad a_k = P_k.$$

These a_i can be shown to be expressible as polynomials in the matrices $\{Z_iZ_i^T\}$ since the P_i can be so expressed.

It follows that P_i and a_i, for all i, are members of $\mathscr{U}$. One also can verify that $\mathscr{U}$ has an identity element. Next, let $\mathscr{U}_1$ be the associative algebra generated by $\mathscr{U}$ and I_m. One checks that this algebra is just $\mathscr{U} \oplus \Re I_m$. We can also see that a_i, for all i, commutes with every $u \in \mathscr{U}_1 = \mathscr{U} \oplus \Re I_m$. Then since the a_i sum to I_m, they give a direct sum decomposition of $\mathscr{U}_1$ into associative ideals of $\mathscr{U}_1$:

$$\mathscr{U}_1 = \mathscr{U}_1 a_1 \oplus \ldots \oplus \mathscr{U}_1 a_k = a_1 \mathscr{U} \oplus \ldots \oplus a_k \mathscr{U}_1$$

Mann [1960; p. 3] correctly reminds us, though, that these ideals are **not** necessarily simple ideals, additional work being required to further decompose them where needed.

The importance of Mann's construction is this: **every family of nested linear hypothesis tests in the standard linear model corresponds to such a direct sum decomposition of an (associative) algebra, and to such**

an idempotent resolution of the identity. In a somewhat similar study of balanced incomplete block models, James [1957] introduced the term **relationship algebra** to describe this generated algebra $\mathcal{U}_1$, again finding that ring-theory ideal structure was a key to finding the needed sums of squares for an analysis of variance.

Observe that the entire analysis and decomposisiton carried out above can be conducted entirely within the Jordan algebra generated by the set

$$\{m_i\} = \{X_i X_i^T\} \cup \mathbf{I},$$

that is, by $\mathcal{J}(\mathcal{S})$ for $\mathcal{S} = \mathcal{S}(m_i)$.

In general $\mathcal{J}(\mathcal{S})$ will be properly contained in $\mathcal{U}_1 \cap \mathbf{S}_m \subseteq \mathcal{U}_1$, so that usually less work will be required to identify all the needed sums of squares (random quadratic forms). Our results on statistical independence and χ^2–distributivity in a Jordan algebra context have thus sharpened and simplified the calculations involved in Mann's associative algebra construction.

Moreover, the algebra $\mathcal{J}(\mathcal{S})$ retains an important practical feature of the associative algebra $\mathcal{U}_1$, a feature utilized in the earlier work on relationship algebras and the algebra of linear hypotheses. That is, consider the linear model:

$$y = Z_1 b_1 + Z_2 b_2 + \ldots + Z_k b_k,$$

for

b_i either a fixed vector,
        or random with expectation zero and covariance $\sigma_i^2 \mathbf{I}_{c(i)}$.

Then if we first find the Jordan algebra $\mathcal{J}$ generated by all the $Z_i Z_i^T$, it follows that quadratic forms based on the possible models having b_i fixed or random, can be studied using some subspace $\mathcal{S}^*$ of $\mathcal{S} = \mathcal{S}(\{Z_i Z_i^T\})$, and some subalgebra $\mathcal{J}^*$ of $\mathcal{J} = \mathcal{J}(\mathcal{S})$. Indeed, if the required subalgebra is an ideal $\mathcal{J}^* = \mathcal{I}^* \subseteq \mathcal{J}$ then $\mathcal{I}$ is isomorphic to the factor algebra $\mathcal{J}/\mathcal{I}$, where $\mathcal{I}$ is the Jordan ideal complement of $\mathcal{I}^*$ in $\mathcal{J}$, with $\mathcal{J} = \mathcal{I} \oplus \mathcal{I}^*$. That such an $\mathcal{I}$ always exists, is called **complete reducibility**, and is a consequence of the **semisimplicity** of $\mathcal{J}$. The definitions of these terms and the verification of our direct–sum result can be found in Curtis and Reiner [1966; p. 39] and Jacobson [1968; Corollary, p. 286]; see also Chapter 3 below.

Also, for such $\mathcal{J}^*$, and a, b $\in \mathcal{J}^*$, the forms $y^T a y$, $y^T b y$ are then independent for all $\mathcal{V} \in \mathcal{J}$ if and only if they are independent for all $\mathcal{V} \in \mathcal{J}^*$.

Thus in application of our methods to the mixed linear model we may always usefully begin by assuming that y has **Jordan covariance structure**: the set of allowable covariances $\mathcal{V}$ spans a Jordan algebra $\mathcal{J}(\mathcal{S})$. This is therefore a useful, logical extension of T. W. Anderson's notion of data having **linear covariance structure**: see Anderson [1969; 1970; 1973]. For the variance component problem we may further assume that a set of positive semi-definite allowable $\mathcal{V}$'s forms a basis set for $\mathcal{J}$. These points just made will be pursued in Chapter 4.

Further understanding in our Jordan algebra approach is thus mainly to be obtained from a more refined look at the statistical consequences known inner structure of the simple Jordan algebras given in **Theorem 3.6** which we now know are all realized in real symmetric matrices.

2.10   A Concluding Example.

Malley [1986; pp. 118–128] discusses the application of the above Jordan algebra results for the complete statistical analysis of the **partially balanced incomplete block designs** (PBIB's). Here the model in its most general form involves *four* random components, and so is not a case covered by **Theorem 4.6** above. By using the proper experimental design it is possible to get optimal estimates of every linear function of the variance components, as can be seen by carefully choosing the design parameters for the PBIB, and by examination of the (one dimensional) space of available optimal unbiased estimators. By choosing different settings for the design parameters one can find optimal estimates for each of components seperately. Naturally, this will usually involve obtaining multiple samples, or splitting an existing sample, but for some combinations of the design parameters it may be possible to identify subsets of a single (random) sample such that each subset will generate the best estimate for one of the model components.

Moreover, the use of generally strictly larger associative algebras, via association schemes, is now completely obviated, as is the need to consider only balanced data, or data having well-characterized, inherent symmetries; c.f. Speed [1987], and Dawid [1988].

References for Chapter 1 and 2

Anderson, T. W. [1969]. "Statistical inference for covariance matrices with linear structure." In: **Multivariate Analysis, II**. P. R. Krishnaiah (editor), 55–66. Academic Press, New York.

Anderson, T. W. [1970]. "Estimation of covariance matrices which are linear combinations or whose inverse are linear combinations of given matrices." In: **Essays in Probability and Statistics**. R. C. Bose et al. (editors), 1–24. University of North Carolina Press.

Anderson, T. W. [1973]. "Asymptotically efficient estimation of covariance matrices with linear structure." **Annals of Statistics, 1**, 135–141.

Anderson, T. W. [1984]. **An Introduction to Multivariate Statistical Analysis**, 2nd Edition. Wiley and Sons, New York.

Bose, R. C., and Mesner, D. M. [1959]. "On linear associative algebras corresponding to association schemes of partially balanced designs." **Annals of Math. Stat., 30,** 21–38.

Braun, H. and Koecher, M. [1966]. **Jordan-algebren**. Springer–Verlag, Berlin.

Cochran, W. G. [1934]. "The distribution of quadratic forms in a normal system with applications to analysis of covariance." **Proc. Cambridge Philos. Soc., 30,** 178–191.

Cohn, P. M. [1982]. **Algebra**. Vol. 1, 2nd Edition, Wiley and Sons, New York.

Curtis, C. W., and Reiner, I. [1966]. **Representation Theory of Finite Groups and Associative Algebras**. 2nd Edition. Wiley, New York.

Curtis, C. W., and Reiner, I. [1981]. **Methods of Representation Theory with Applications to Finite Groups and Orders**. Wiley, New York.

Dawid, A. P. [1988]. "Symmetry models and hypotheses for structured data layouts (with discussion)." **J. Roy. Statist. Ser., Series B, 50,** 1–34.

Dembo, A., Mallows, C. L., and L. A. Shepp [1989]. "Embedding nonnegative definite Toeplitz matrices in nonnegative definite circulant matrices, with application to covariance estimation." **IEEE Trans. on Info. Theory, 35(6),** 1206–1212.

Dempster, A. P., Laird, N. M., and D. R. Rubin [1977]. "Maximum likelihood from incomplete data via the EM algorithm (with discussion)." **J. Roy. Statist. Soc., Series B, 39,** 1–38.

Diaconis, P. [1988]. **Group Representations in Probability and Statistics**. IMS Lecture Notes – Monograph Series, Vol. 11. Institute of Mathematical Statistics, Hayward, California.

Eaton, M. L. [1989]. **Group Invariance Applications in Statistics**. Regional Conference in Probability and Statistics, Vol. 1. Institute of Mahematical Statistics (Hayward, California), and the American Statistical Association, Alexandria, Virginia.

Elbassiouni, M. Y. [1983]. "On the existence of explicit restricted maximum likelihood estimators in multivariate normal models." **Sankhya, Series B, 45,** 303–305.

Farrell, R. H. [1985]. **Multivariate Calculation. Use of the Continuous Groups.** Springer–Verlag, New York.

Hannan, E. J. [1970]. **Multiple Time Series.** Wiley and Sons, New York.

Helgason, S. [1984]. **Groups and Geometric Analysis: Integral Geometry, Invariant Differential Operators, and Spherical Functions.** Academic Press, San Diego.

Herstein, I. N. [1964]. **Topics in Algebra.** Xerox College Publishing, Lexington, MA.

Herstein, I. N. [1969]. **Topics in Ring Theory.** University of Chicago Press, Chicago.

Herstein, I. N. [1968]. **Noncommutative Rings.** Carus Mathematical Monograph No. 15, The Mathematical Association of America.

Jacobson, N. [1953]. **Lectures in Abstract Algebra.** Vol. II. D. Van Nostrand, New York.

Jacobson, N. [1968]. **Structure and Representation of Jordan Algebras.** American Mathematical Society, Rhode Island.

Jacobson, N. [1987]. "Jordan algebras of real symmetric matrices." **Algebras, Groups and Geometries, 4,** 291–304.

James, A. T. [1957]. "The relationship algebra of an experimental design." **Annals of Math. Stat., 28,** 993–1002.

Jensen, S. T. [1988]. "Covariance hypotheses which are linear in both the covariance and inverse covariance." **Annals of Statistics, 16,** 302–322.

Jordan, P., von Neumann, J., and Wigner, E. [1934]. "On an algebraic generalization of the quantum mechanical formalism." **Annals of Math., 35,** 29–64.

Malley, J. D. [1986]. **Optimal Unbiased Estimation of Variance Components.** Lecture Notes in Statistics, Vol. 39, Springer–Verlag, New York.

Malley, J. D. [1987]. "Subspaces and Jordan algebras of real symmetric matrices." **Algebras, Groups and Geometries, 4,** 265–289.

Mann, H. B. [1960]. "The algebra of a linear hypothesis." **Annals of Math. Stat., 31,** 1–15.

McCoy, N. H. [1964]. **The Theory of Rings**. Macmillan, Toronto, Ontario.

McCrimmon, K. [1969]. "On Herstein's theorems relating Jordan and associative algebras." **Journal of Algebra, 13,** 382–392.

McCrimmon, K. [1969]. "Nondegenerate Jordan rings are von Neumann regular." **Journal of Algebra, 11,** 111–115.

McCrimmon, K. [1984]. "The Russian revolution in Jordan algebras." **Algebras, Groups and Geometries, 1,** 1–61.

Ogawa, J., and Ishii, G. [1965]. "The relationship algebra and the analysis of variance of a partially balanced incomplete block design." **Annals of Math. Stat., 36,** 1815–1828.

Paige, L. J. [1963]. "Jordan algebras." In: **Studies in Modern Algebra**, A. A. Albert (editor). Prentice-Hall, New Jersey. 144–186.

Pukelsheim, F. [1981]. "On the existence of unbiased nonnegative estimates of variance covariance components." **Annals of Math. Stat., 9,** 293–299.

Rao, C.R., and Mitra, S.K. [1971]. **Generalized Inverse of Matrices and its Applications**. Wiley, New York.

Robinson, J. [1970]. "On relationship algebras of incomplete block designs." **Annals of Math. Stat., 41,** 648–650.

Robinson, J. [1971]. "Correction to "On relationship algebras of incomplete block design." **Annals of Math. Stat., 42,** 842.

Sagle, A. A., and Walde, R. E. [1973]. **Introduction to Lie Groups and Lie Algebras**. Academic Press, San Diego.

Schafer, R. D. [1966]. **An Introduction to Nonassociative Algebras.** Academic Press, New York.

Seber, G. A. F. [1980]. **The Linear Hypothesis: a General Theory.** 2nd Edition. Macmillan, New York.

Seely, J. [1971]. "Quadratic subspaces and completeness." **Annals of Math. Stat., 42,** 710–721.

Speed, T. P. [1987]. "What is an analysis of variance? (Special Invited Paper, with Discussion)." **Annals of Statistics, 15,** 885–941.

Szatrowski, T. H. [1980]. "Necessary and sufficient conditions for explicit solutions in the multivariate normal estimation problem for patterned means and covariances." **Annals of Statistics, 11**, 802–810.

Szatrowski, T. H., and Miller, J. J. [1980]. "Explicit maximimum likelihood estimates from balanced data in the mixed model of the analysis of variance." **Annals of Statistics, 8**, 811–819.

Upmeier, H. [1987]. **Jordan Algebras in Analysis, Operator Theory, and Quantum Mechanics**. Regional Conference Series in Mathematics, Vol. 67. American Mathematical Society, Providence, Rhode Island.

van der Waerden, B. L. [1970]. **Algebra**. 7th Edition, Frederick Ungar, New York.

Wijsman, R. A. [1990]. **Invariant Measures on Groups and Their Use in Statistics**. IMS Lecture Notes – Monograph Series, Vol. 14. Institute of Mathematical Statistics, Hayward, California.

Chapter 3   Further Technical Results on Jordan Algebras

3.0 Outline of this Chapter.

In this chapter we fill in details of some of the proofs given in Chapter 2. Also, new results are presented about subspaces and ideals in Jordan algebras that should help further the application of Jordan algebras in statistical problems generally. To assist in this latter task, our treatment of the material is designed to make the chapter largely self-contained, hence many of the definitions and first appearing in Chapter 2 are re-introduced, though more compactly. This chapter is unabashedly much more mathematical than any of the others, and unavoidably so, since the complete proofs of many of the results (previously just stated) are comparatively non-trivial. Many, but not all, of the results appearing here first appeared in Malley [1987].

We begin with a more complete accounting of the invention of Jordan algebras.

In 1934 Jordan, von Neumann and Wigner introduced the idea of an "r-number algebra," the algebra we now call a Jordan algebra, and first referred to as such by A. A. Albert in 1946. This r-number algebra was introduced in an effort to axiomatize the foundations of the then-new quantum mechanics. While this first definition required that the algebra be *finite* dimen-sional, and was thus ultimately unsuitable for many of even the simplest important physical realizations, the notion has still inspired an enormous mathematical literature.

The standard modern references are Jacobson [1968], and Braun and Koecher [1966]. Also useful is Schafer [1967], on general non-associative algebras, along with Emch [1972] that integrates the algebra, statistics and physics involved in Jordan algebras in their original context as well as in the later quantum field theory. Additional background material on abstract algebra will be found in Lang [1967], van der Waerden [1970], and Greub [1978].

As part of their early pioneering work on Jordan algebras, Jordan von Neumann and Wigner also completely characterized all formally real, finite dimensional Jordan algebras. This early great effort will referred to as simply JNW, or, the JNW theorem.

Moving in a completely different realm, the author's researches on optimal statistical estimation (Malley [1986]), lead to a need for a complete classification of all finitely generated Jordan algebras of real symmetric matrices. Since all these special Jordan algebras are easily shown to be formally real semisimple (definitions below), the JNW theorem applies. The only obstacles then to a complete classification are the exclusion of all *exceptional* (that is, non-special) algebras appearing in JWN, and the demonstration that the remaining cases are indeed all explicitly realized as finitely generated algebras of real symmetric matrices. This classification then amounts essentially to a concluding refinement of JNW.

The first problem is solved here immediately, while a solution of the second was first solved for this author by Nathan Jacobson (and nearly instaneously, once the question was first raised, evidently for the time, by Malley). Our task is to discuss these solutions and present our own solution of the second. See also Jacobson [1987] where full details of his solution of the second problem are given. As in part we prove the same result, there is necessarily some overlap, especially in the introductory material of our Section 1.

Also presented are new and, we hope, useful results about the subspace and ideal structure of real Jordan matrix algebras that are, in most cases, valid for any finite dimensional special Jordan algebra. The chapter concludes with an application to the equations that arose in the same statistical context that lead us to our original classification problem, namely the necessary and sufficients conditions for statistical independence and χ^2–distributivity contained in Chapter 2.

Finally, because of the higher mathematical density of this chapter, we explicitly number many of the equations, and also (in this chapter only) use the notation " ' " for matrix traspose.

3.1 The JNW Theorem.

Recall from Chapter 2 that a Jordan algebra $\mathcal{J}$, for our purposes, an algebra over $\Re$ with product " . " such that

a. b = b . a,   and $(a^{.2} . b) . a = a^{.2} . (b . a)$,   where $a^{.2} = a . a$.

The algebra $\mathcal{J}$ is called **formally real** if

(1.1) $a^{.2} + b^{.2} = 0 \Rightarrow a = b = 0$, for all a, b $\in \mathcal{J}$.

Consider next an associative algebra $\mathcal{A}$ with product ab. Write $\mathcal{A}^+$ for the algebra with the same elements and new product:

(1.2) $a . b = \frac{1}{2}(ab + ba)$ for all a, b $\in \mathcal{A}$.

One checks that $\mathcal{A}^+$ is Jordan, with $a^{.n} = a^n$. Thus the concepts of nilpotence ($a^n = 0$) and idempotence ($a^2 = a$) are the same for $\mathcal{A}$ and $\mathcal{A}^+$. An important role is played by the **Jordan triple product**:

(1.3)   {abc} = a . (b . c) + (a . b) . c − (a . c) . b .

The algebra $\mathcal{J}$ is **special** if there if there is an algebra monomorphism (one to one linear mapping that respects the algebra product) of $\mathcal{J}$ into $\mathcal{A}^+$, for some associative algebra $\mathscr{A}$. Otherwise it is **exceptional**. Thus special Jordan algebras are just subspaces of associative algebras $\mathscr{A}$ that are closed under the product a . b. If $\mathscr{J}$ contains the unit (identity element) 1 of $\mathscr{A}$, this can be shown to be equivalent to closure under the triple product {abc}, since then

   a . b = {ab1}.

Recall that the canonical example of a special, formally real Jordan algebra is $\mathscr{J}$ = $\mathbf{S}_m$, the algebra of all real symmetric matrices in the associative algebra

   $\mathscr{A} = [\mathfrak{R}]_m$ of all real $m \times m$ matrices.

Next define the Jordan algebra of a given symmetric bilinear form f(x,y) on a vector space $\mathscr{B}$. This is given by $\mathscr{J} = \mathfrak{R} \oplus \mathscr{B}$, where addition is component-wise and the product is

$$(\alpha 1 + x) . (\beta 1 + y) = [\alpha\beta + f(x,y)]1 + [\beta x + \alpha y],$$
$$\text{for all } a, b \in \mathfrak{R}, \text{ and } x, y \in \mathscr{B}.$$

Further, for $\mathscr{J}$ finite dimensional over $\mathfrak{R}$, the **radical** of $\mathscr{J}$, written rad $\mathscr{J}$, is its maximal nilpotent ideal, where an ideal is **nilpotent** if for some integer N every product (in any association or grouping by parentheses) of N elements is zero. The algebra $\mathscr{J}$ is declared to be **semisimple** if its radical is zero, and **simple** if it has no proper non-zero ideals, and $\mathscr{J}^2 \neq 0$. For example, a formally real algebra is always semisimple since it cannot even contain a non-zero nilpotent element.

One more item of notion is needed before stating the fundamental JNW result. Thus, suppose the algebra $\mathscr{D}$, not necessarily associative, has an involution * :

   a → a* ,   for all a ∈ $\mathscr{D}$,

with

   (a + b)* = a* + b*,   a** = a,   and   (ab)* = b*a*,   for all a, b ∈ $\mathscr{D}$.

Consider $[\mathscr{D}]_m$, the algebra of $m \times m$ matrices over $\mathscr{D}$. Then * induces a canonical involution on $[\mathscr{D}]_m$ by

   $[d_{ij}] \rightarrow \{(d_{ij})^*]'$

for [•]' the transpose of matrix [•]. For $\mathscr{D} = \mathfrak{R}$, the involution induces the transpose on $[\mathfrak{R}]_m$, and $\mathbf{S}_m$ is exactly the algebra of matrices that are transpose-invariant. Similarly for the complex numbers, C, the usual conjugation

   a + bi → a − bi,

induces the conjugate transpose involution on $[C]_m$. For Q, the real quaternions, * is given by:

$$\alpha + \beta i + \gamma j + \delta k \ \rightarrow \ \alpha - \beta i - \gamma j - \delta k .$$

Finally, the non-associative, noncommutative algebra of octonions O is a featured player in the complete structure theory of Jordan algebras; for its definition, and involution, consult Jacobson [1968, p. 17].

We can now state the basic:

1.4 Theorem JNW. (Jordan, von Neumann, and Wigner [1934]). Every finite dimensional, formally real Jordan is semisimple, has an identity element, and is a direct sum of simple ideals. Each of these simple ideals is isomorphic to one of the following: **(1)** $\Re$; **(2)** the Jordan algebra $\mathcal{J} = \Re \oplus \mathcal{B}$ of a positive definite symmetric bilinear form f(x,y) on $\mathcal{B}$; **(3)** the algebra of elements of $[\mathcal{D}]_n$, for $\mathcal{D} = \Re$, C (the complex numbers), or Q, the real quaternions, which are left fixed by the conjugate transpose involution induced by the usual conjugation on $\Re$, C or Q, or, if n = 3, $\mathcal{D} = O$, the octonion algebra over $\Re$ with its standard conjugation.

Let's recall two standard Jordan facts: first, that $\mathcal{J} = \Re \oplus \mathcal{B}$ is indeed special (Jacobson [1968; p. 261]), and second, that the matrices, over the octonions, which are fixed by the conjugate transpose involution, are an exceptional Jordan algebra (Jacobson [1968; Chapter IX). It then follows that the assumptions of JNW, along with $\mathcal{J}$ special allow us to obtain a first classification for such $\mathcal{J}$ by simply excluding the octonion case in the list above, and as an immediate application of this we get:

1.5 Theorem. Suppose $\mathcal{J}$ is the Jordan algebra generated by a set of real symmetric matrices. Then $\mathcal{J}$ is formally real, and therefore is semisimple, has an identity element, and is a direct sum of simple ideals. These simple ideal components are among the types **(1)**, **(2)**, and **(3)** above obtained after deleting the matrix algebra of symmetric elements over the octonions.

Proof. We need only show that $\mathbf{S}_m$ is formally real since then any subalgebra will be formally real, and $\mathcal{J} \subseteq \mathbf{S}_m$. Hence, as observed above, $\mathcal{J}$ must be semisimple.

Thus, suppose a, b $\in \mathbf{S}_m$, for m the order of the generating matrices, and observe that for any d $\in [\Re]_m$,

$$\text{tr} (dd') \ = \ \Sigma\, \Sigma\ (d_{ij})^2 ,$$

so tr(dd') $\geq$ 0, and = 0 if and only if d = 0. From this positivity of the trace we have the basic fact, which we later use repeatedly,

(1.6 Lemma.) dd' = 0 for d $\in [\Re]_m$ $\Leftrightarrow$ tr(dd') = 0 $\Leftrightarrow$ d = 0, as well as:

(1.7 Lemma.) $a^2 + b^2 = a^2 + b^2 = 0$

Proof. We have

$$0 = \text{tr}(a^2 + b^2) = \text{tr}(a^2) + \text{tr}(b^2) = \text{tr}(aa') + \text{tr}(bb')$$

$$\Leftrightarrow \quad a = b = 0 \ .$$

Hence $\mathbf{S}_m$ is formally real, and the result follows from the JNW. ∎

The Theorem also appears in Braun and Koecher [1966], as Theorem 5.6, p. 331. Alternatively a proof can also be found by using Exercise 6, p. 211 of Jacobson [1968] along with Theorem 11, p. 210. These lead to a complete classification of all finite dimensional, special, central simple, Jordan algebras, so it is only then required to exclude from the list of Exercise 6 all those cases that end with the Jordan algebra not formally real.

With this result the first problem mentioned in the introduction is solved, and we turn now to the second problem, that of the verification that the three algebra types occurring in JNW, after excluding the octonion case, are in fact all realized for specific sets of symmetric matrices.

3.2 The Classes of Simple Formally Real, Special Jordan Algebras.

The difficult case is that of the Jordan algebra $\mathscr{J}$ of a real positive definite symmetric bilinear form, and this is handled with:

2.1 Theorem. The Jordan algebra $\mathscr{J}$ of a positive definite symmetric bilinear form f on a vector space $\mathscr{B}$ of dimension $k > 1$ is isomorphic to a Jordan algebra $\mathscr{J}^*$ of real symmetric matrices, $\mathscr{J} \approx \mathscr{J}^* \subseteq \mathbf{S}_m$ for some m, as shown by:

   (i) $\mathscr{J} \subseteq \mathscr{C}^+$, for $\mathscr{C}$ the Clifford algebra $\mathscr{C}$ of f;
   (ii)   the left regular representation of $\mathscr{C}$ is a faithful unital representation of $\mathscr{C}$ on a vector space $\mathscr{V} = \mathscr{C}$ of dimension $n = 2^k$;
   (iii)   every unital representation R of $\mathscr{C}$ on a vector space $\mathscr{V}$ has

$$R(\mathscr{J}) \subseteq \mathscr{S}(\mathscr{V}),$$

   for $\mathscr{S}(\mathscr{V})$ the space of self-adjoint transformations on $\mathscr{V}$, relative to some positive definite inner product $(\,\bullet\,,\,\bullet\,)_{\mathscr{V}}$ on $\mathscr{V}$; hence $R(\mathscr{J}) \subseteq \mathbf{S}_m$ is a Jordan algebra of real symmetric matrices with respect to an orthonormal basis for $\mathscr{V}$;
   (iv)   since $k > 1$ every unital representation R of $\mathscr{C}$ is faithful on $\mathscr{J}$.

Proof. Our references for Clifford algebras are Greub [1978; Chapters 10 and 11] and Jacobson [1968; §7.1].

Start by recalling that $\mathscr{C}$, the Clifford algebra over the inner product space $\mathscr{B}$, is defined as $\mathscr{T}(\mathscr{B})/\mathscr{K}$, where $\mathscr{T}(\mathscr{B})$ is the tensor algebra of $\mathscr{B}$, and $\mathscr{K}$ is the ideal of $\mathscr{T}(\mathscr{B})$ generated by all elements of the form $\{x \otimes x - f(x,x)1\}$. We are free to choose an orthonormal basis $\{e_i\}$ for the positive definite form f, so

$$f(e_i , e_j) = \delta_{ij} .$$

Then in $\mathscr{C}$ we have

(2.2) $(e_i)^2 = 1 , \quad e_i e_j = - e_j e_i ,$

(2.3) 1 and all the products $e_{i_1} e_{i_2} \ldots e_{i_n}$ $(i_1 < i_2 < \ldots < i_n)$ form a basis for $\mathscr{C}$ and it follows that the canonical injection of $\mathscr{B}$ in $\mathscr{C}$ can be extended to $\mathscr{J}$ in $\mathscr{C}$, as in (i) of the Theorem. This is essentially Theorem 1 of Jacobson [1968; p. 261], which shows that $\mathscr{C}$ is a unital universal special (associative) envelope for $\mathscr{J}$.

Since $\mathscr{C}$ is unital, its left regular representation with $R(c) = c_l$ = left multiplication by c, is faithful, as in (ii).

For (iii), begin by letting let R be any unital representation of $\mathscr{C}$ in $\mathrm{End}(\mathscr{V})$ for some vector space $\mathscr{V}$, so $R(1) = 1_{\mathscr{V}} \in \mathrm{End}(\mathscr{V})$.

From (2.2) we first see that the elements e_i generate a finite group $\mathscr{G}$ contained in $\mathscr{C}^*$, the multiplicative group of invertible elements of $\mathscr{C}$. In fact, $\mathscr{G}$ consists precisely of the 2^{k+1} elements $(\pm c)$ for the 2^k basis elements c of (2.3). Pick arbitrarily a positive definite inner product $(\bullet , \bullet)_0$ on $\mathscr{V} \cong \mathfrak{R}^n$, for example $(a')(b)$ for a, b column vectors in $\mathfrak{R}^n$, and use a standard hermitian argument of averaging the inner product to get a new one:

$$(u,v)_{\mathscr{V}} = \sum_{g \in \mathscr{G}} (R(g)u, R(g)v)_0 .$$

This is seen to be positive definite again, but also now has each $R(h)$ for $h \in \mathscr{G}$, acting orthogonally by definition:

$$
\begin{aligned}
(R(h)u, R(h)v)_{\mathscr{V}} &= \sum_{g \in \mathscr{G}} (R(g)R(h)u, R(g)R(h)v)_0 \\
&= \sum_{g \in \mathscr{G}} (R(gh)u, R(gh)v)_0 \\
&= \sum_{g' \in \mathscr{G}} (R(g')u, R(g')v)_0 \\
&= (u,v)_{\mathscr{V}} .
\end{aligned}
$$

Next, since $(e_i)^2 = 1$ it follows that $R(e_i)^2 = 1$, so $R(e_i)$ is invertible. Hence for $h = e_i \in \mathscr{G}$ we have $R(e_i)^-$, the adjoint of $R(e_i)$ relative to

$$(\bullet , \bullet)_{\mathcal{V}} = R(e_i)^{-1} = R(e_i),$$

and certainly $R(1)^{\sim} = R(1) = 1_{\mathcal{V}}$. Thus $R(x)^{\sim} = R(x)$ for all basis elements of $\mathcal{J}$, so by linearity, all $R(x)$ for $x \in \mathcal{J}$, are self-adjoint and $R(x) \in \mathcal{S}(\mathcal{V})$. Choosing any orthonormal basis $\{v_i\}$ for $\mathcal{V}$, relative to $(\bullet , \bullet)_{\mathcal{V}}$, we thus see

$$(R(x)v_i, v_j)_{\mathcal{V}} = (v_i, R(x)v_j)_{\mathcal{V}}, \quad \text{for all } i, j$$

so that $R(x) \subseteq \mathbf{S_m}$, and this completes the verification of **(iii)**.

Finally, to check faithfulness as in **(iv)**, we actually only need nondegeneracy of f, not positive definiteness. For then one can quickly show that $\mathcal{J}$ is simple when $\dim \mathcal{B} > 1$, and since R is also a unital Jordan homomorphism,

$$\mathcal{J} \to \{\text{End}(\mathcal{V})\}^+,$$

of Jordan algebras we have $\mathcal{J} \cap \ker R \neq \mathcal{J}$. By simplicity then,

$$\mathcal{J} \cap \ker R = 0$$

and R is faithful on $\mathcal{J}$.

We can also show directly that any unital Jordan homomorphism H,

$$H : \mathcal{J} \to \mathcal{A}^+,$$

is faithful. Thus if

$$H(\alpha 1 - x) = 0 \quad \text{for } \alpha \in \mathfrak{R}, \quad x \in \mathcal{B},$$

then we may assume that $\alpha = 1$ or 0. Next, from

$$0 = H(\alpha 1 - x) . H(y)$$

$$= H((\alpha 1 - x) . y) = H(\alpha y - f(x,y)1)$$

$$= \alpha H(y) - f(x,y)1_{\mathcal{V}},$$

we have

$$\alpha H(y) = f(x,y) 1_{\mathcal{V}},$$

for all $y \in \mathcal{B}$. If $\alpha = 0$ then $f(x,y) = 0$ for all $y \in \mathcal{B}$, so by non-degeneracy only $x = 0$ is possible. If $\alpha = 1$ then $H(y) = f(x,y)1_{\mathcal{V}}$ for all $y \in \mathcal{B}$, and therefore

$$f(x,y) f(x,z) 1_{\mathcal{V}} = H(y) . H(z)$$

$$= H(y \cdot z) = H(f(y,z)1)$$

$$= f(y,z)1_y \, ,$$

for all $y, z \in \mathcal{B}$. But this is impossible if $\dim \mathcal{B} > 1$, since then there must exist $y \neq x$ with $f(x,y) = 0$, so $f(y,z) = f(x,y)f(x,z) = 0$, for all $z \in \mathcal{B}$, and by nondegeneracy this is a contradiction.

Thus $H(\alpha 1 - x) = 0$ can occur only if $\alpha = x = 0$, and the proof is complete. ∎

Example.

To illustrate the above imbedding process, consider the generalized quaternion algebra $\mathcal{A} = [\mathfrak{R}]_2$, with basis elements $\{1, a, b, ab\}$, where $a^2 = b^2 = 1$, and $ab + ba = 0$ (where $a = e_{11} - e_{22}$, $b = e_{12} + e_{21}$). Let

$$(c, d) = \tfrac{1}{2} \operatorname{tr}(c'd) \quad \text{for } c, d \in [\mathfrak{R}]_2 \, .$$

so that

$$(a,a)_0 = (b,b)_0 = 1 \, , \quad (a,b)_0 = 0.$$

Using the vec operator we identify $\mathcal{A}$ with $\mathfrak{R}^4$:

$$c = (c_{ij}) \in \mathcal{A}, \; c \to \operatorname{vec}(c) = (c_{11}, c_{21}, c_{12}, c_{22})' \in \mathfrak{R}^4.$$

We have $\operatorname{tr}(c'd) = (\operatorname{vec} c')' (\operatorname{vec} d) = $ the usual inner product in $\mathfrak{R}^4$.

Let space $\mathcal{B}$ be the span of $\{a, b\}$, and $\mathcal{J}$ the span of $\{1, a, b\}$. Then $\mathcal{J}$ is the Jordan algebra of $\mathcal{B}$ relative to $(\bullet , \bullet)_0$, $\mathcal{A}$ is a Clifford algebra and the left regular representation of $\mathcal{A}$ sends $1 \to 1$, and

$$a \to \begin{bmatrix} 0 & 1 & 0 & 0 \\ 1 & 0 & 0 & 0 \\ 0 & 0 & 0 & 1 \\ 0 & 0 & 1 & 0 \end{bmatrix} \qquad b \to \begin{bmatrix} 0 & 0 & 1 & 0 \\ 0 & 0 & 0 & -1 \\ 1 & 0 & 0 & 0 \\ 0 & -1 & 0 & 0 \end{bmatrix}$$

$$ab \to \begin{bmatrix} 0 & 0 & 0 & -1 \\ 0 & 0 & 1 & 0 \\ 0 & -1 & 0 & 0 \\ 1 & 0 & 0 & 0 \end{bmatrix}$$

Hence 1, a and b, but **not** ab map into $\mathbf{S}_4$, so $\mathcal{J} \subseteq \mathbf{S}_4$. Note that already $\mathcal{J} \subseteq \mathbf{S}_2$ in terms of the identification $\mathcal{A} = [\mathfrak{R}]_2$.

If instead of the usual inner product on $\mathcal{V} = \mathcal{A} = \mathfrak{R}^4$ we take the hermitian averaged product $(\cdot , \cdot)_{\nu}$ of the theorem, we find:

$$(u,v)_{\nu} = [(u,v)_0 + (a(u), a(v))_0 + (b(u), b(v))_0 + (ab(u), ab(v))_0]$$

$$= \tfrac{1}{2} \operatorname{tr} [u'(1 + a'a + b' b + (ab)'(ab))v]$$

$$= \tfrac{1}{2} \operatorname{tr} (u'(4)v) = 4(u,v)_0 .$$

So $R(\mathcal{J}) \subseteq \mathcal{S}(\mathcal{V})$ relative to $(\cdot , \cdot)_{\nu}$ as required.

Also, any generalized quaternion algebra is by definition of the form

$$\{1, a, b, ab\} \quad \text{with} \quad a^2 = \lambda 1 , \quad b^2 = \mu 1, \quad ab + ba = 0 , \quad \text{for } \lambda, \mu \in \mathfrak{R},$$

and these are all Clifford algebras: for the real quaternions $\mathcal{Q}$, the space $\mathcal{B}$ is that spanned over $\mathfrak{R}$ by $a = i(e_{11} - e_{22})$, $b = e_{12} - e_{21}$, for $a, b \in [C]_2$, with bilinear form $f(x,y) = \tfrac{1}{2} (xy^* + yx^*)$. $\square$

We state our main classification theorem:

2.5 Theorem.   A finite dimensional special Jordan algebra $\mathcal{J}$ is formally real if and only if it is isomorphic to a finitely generated Jordan algebra of real symmetric matrices, and is isomorphic to a direct sum of one of the following:
(1) $\mathfrak{R}$; **(2)** the Jordan algebra $\mathcal{J} = \mathfrak{R} \oplus \mathcal{B}$ of a positive definite symmetric form on a vector space $\mathcal{B}$; **(3)** the Jordan algebra of symmetric elements of $[\mathcal{A}]_n$, where $\mathcal{A} = \mathfrak{R}, C$ or $\mathcal{Q}$, the real quaternions, and symmetry is with respect to the involutions: transpose, complex conjugate transpose, or quaternion conjugate transpose, respectively.

Proof.   Given **Theorem 1.5** and the JNW, **Theorem 1.4**, we need only show that the three cases of the theorem as stated can each be imbedded in $\mathbf{S}_m$ for some m.

However, the case $\mathcal{J} = \mathfrak{R} \oplus \mathcal{B}$ is our theorem proved above, while the case $\mathcal{A} = \mathfrak{R}$ is immediate. The case $\mathcal{A} = C$ follows from the standard left regular injection of C in $[\mathfrak{R}]_2$, where one checks then that the map does respect the involutions. The case for $\mathcal{A} = \mathcal{Q}$ is similar, using the standard left regular injection of $\mathcal{Q}$ in $[\mathfrak{R}]_4$: see for example Herstein [1966; p. 8]. ∎

Jacobson was first to publish a derivation of the injection of $\mathcal{J}$ in $\mathbf{S}_m$ for $\mathcal{J} = \mathfrak{R} \oplus \mathcal{B}$, with f a positive definite symmetric form on $\mathcal{B}$. The central idea is use of the known, fully elaborated structure of the Clifford algebra for any $\mathcal{B}$ with non-degenerate form f, and as might be expected, this additional knowledge provides much more precise information about the imbedding. We also note that Jensen [1988] gave a derivation that results in the same classification presented here, hence one less precise than that of Jacobson. Moreover, the methods used were more involved than either ours or Jacobson's as it invoked both the Clifford structure theory and our hermitian argument.

3.3. The Jordan and Associative Closures of Subsets of S_m.

We now examine more closely the Jordan and associative algebras generated by a set of matrices in S_m. By finite dimensionality we may as usual assume the set is finite, and write $\{m_i\}$, $1 \le i \le n$, for a set of matrices in S_m, $\mathscr{S} = \mathscr{S}\{m_i\}$ for the vector space they span in S_m. Then:

3.1 Theorem. Let $\{m_i\} \subseteq S_m$ be a set of symmetric matrices, $\mathscr{S}$ the vector space they span, $\mathscr{J}$ the Jordan subalgebra they generate in S_m, and $\mathscr{A}$ the associative subalgebra they generate in $[\mathfrak{R}]_m$. Then

(i) $\mathscr{A}$ is semisimple and invariant under the transpose involution;

(ii) $\mathscr{A} \cap S_m$ is a formally real, semisimple Jordan algebra generated by $\{m_i\}$ and all *tetrads* of the form

$$m_{i_1} m_{i_2} m_{i_3} m_{i_4} + m_{i_4} m_{i_3} m_{i_2} m_{i_1} \quad \text{for } i_1 < i_2 < i_3 < i_4$$

(so $\mathscr{A} \cap S_m = \mathscr{J}$ if there are ≤ 3 generators m_i);

(iii) $\mathscr{J}$ is a formally real, semisimple Jordan algebra;

(iv) $\mathscr{J}$ can be recursively constructed through the chain
$$\mathscr{S} = \mathscr{S}_0 \subseteq \mathscr{S}_0 \subseteq \dots \subseteq \mathscr{S}_p = \mathscr{S}_{p+1} = \dots = \mathscr{J},$$

where

$$\mathscr{S}_{i+1} = \mathscr{S}_i + \left\{ \sum t_{jk}\left(b_j b_k + b_k b_j\right) \middle| \text{ all } t_{jk} \in \mathfrak{R}, \text{ and all } b_j, b_k \in \mathscr{S}_i \right\};$$

(v) if $\mathscr{S}$ contains an identity element e, $es = se = s$ for all $s \in \mathscr{S}$, then $\mathscr{J}$ can be constructed through

$$\mathscr{S} = \mathscr{T}_0 \subseteq \mathscr{T}_1 \subseteq \dots \subseteq \mathscr{T}_q = \mathscr{T}_{q+1} = \mathscr{J},$$

where

$$\mathscr{T}_{i+1} = \{ \sum t_{jk} b_j b_k b_j \mid \text{ all } t_{jk} \in \mathfrak{R}, \text{ and all } b_j, b_k \in \mathscr{T}_i \}$$

(vi) for $\mathscr{T}_i$, $\mathscr{T}_i$ as above, we have $\mathscr{S}_i \subseteq \mathscr{T}_i \subseteq \mathscr{S}_{2i}$, and $q \le p \le 2q$.

Proof. For (i), $\mathscr{A}$ is seen to be transpose-invariant since its generators m_i are such. $\mathscr{A}$ is in fact simply the span of all monomials in the m_i, and on these transpose acts by reversal since $m_i \in S_m$. But any transpose-closed subalgebra $\mathscr{A}$ of $[\mathfrak{R}]_m$ is necessarily semisimple: if $a \ne 0$, $a \in \text{rad}\,\mathscr{A}$, then aa' is symmetric, non-zero by **Lemma 1.6**, and also in the ideal $\text{rad}\,\mathscr{A}$, since $a' \in \mathscr{A}$, yet we have seen from **Lemma 1.7** that S_m contains no non-zero nilpotent elements.

For **(ii)**, any subalgebra $\mathcal{J}$ of $\mathbf{S}_m$ is again formally real, and hence as above, or by **Theorem 1.5**, is semisimple. Further, $\mathcal{A} \cap \mathbf{S}_m$ is spanned by the k-tads

$$m_{i_1} \cdots m_{i_k} + m_{i_k} \cdots m_{i_1},$$

while a result of Cohn's (see Jacobson [1968, p. 8]) says these are generated as Jordan products $a \cdot b$ out of all the m_i and all the tetrads.

Part **(iii)** follows directly from **Theorem 1.5**.

Next, let's consider **(iv)**. By construction $\mathcal{S}_i \subseteq \mathcal{S}_{i+1} \subseteq \mathcal{J}$, so by finite dimensionality the chain becomes stationary: $\mathcal{S}_p = \mathcal{S}_{p+1} = \ldots \subseteq \mathcal{J}$. But then $\mathcal{S}_p$ must be closed under the Jordan product $a \cdot b$, since $a, b \in \mathcal{S}_p$ implies

$$ab + ba \in \mathcal{S}_{p+1} = \mathcal{S}_p.$$

Hence $\mathcal{S}_p$ is a Jordan subalgebra containing $\mathcal{S} = \mathcal{S}_0$, therefore containing the smallest such subalgebra $\mathcal{J}$, implying $\mathcal{S}_p = \mathcal{J}$.

Part **(v)** follows from **(iv)**, since as soon as $\mathcal{T}_q$ is closed under triple products it is also closed under Jordan products, as $a \cdot b = \{abe\}$.

Finally for **(vi)**, we verify the inclusions $\mathcal{S}_i \subseteq \mathcal{T}_i \subseteq \mathcal{S}_{2i}$. The first holds by induction because $a \cdot b = \{abe\}$, and the second follows by induction using

$$aba = 2(b \cdot a) \cdot a - b \cdot (a^2).$$

That $q \le p \le 2q$ is now immediate and this completes the proof. ∎

3.4 Subspaces and Ideals in $\mathcal{S}$.

In certain statistical situations we are concerned with subspaces and ideals of $\mathcal{J}$ that are contained in $\mathcal{S}$. The results presented here in fact work quite generally for all special Jordan algebras, except as noted, when we specialize to formally real special Jordan algebras, that is, subspaces and algebras of real symmetric matrices. Assume throughout that the base field (field of coefficients) for $\mathcal{J}$ has characteristic $\ne 2$, so $\tfrac{1}{2} \in \mathcal{J}$.

4.1 Theorem.   Let $\{m_i\}$ be a set of generators for a special Jordan algebra $\mathcal{J} \subseteq \mathcal{A}^+$. Assume the linear span $\mathcal{S}$ of the $\{m_i\}$ has an identity element e, $es = se = s$ for all $s \in \mathcal{S}$. Then

  (i) $\mathcal{B}$ is an ideal of $\mathcal{J}$ if and only if it is an additive subgroup closed under the triple products sbs, for all $s \in \mathcal{S}$, $b \in \mathcal{B}$; equivalently if and only if

$$\{m_i b m_j\} \in \mathcal{B} \text{ for all } i, j \text{ and } b \in \mathcal{B};$$

  (ii) the maximal ideal $\mathcal{B}_.$ of $\mathcal{J}$ contained in $\mathcal{S}$ can be constructed through the chain

$$\mathscr{S} = \mathscr{R}_0 \supseteq \mathscr{R}_1 \supseteq \mathscr{B}_2 \supseteq \ldots \supseteq \mathscr{B}_t = \mathscr{B}_{t+1} = \ldots = \mathscr{B}_* ,$$

where

$$\mathscr{B}_{i+1} = \left\{ b \in \mathscr{B}_i \mid \; sbs \in \mathscr{B}_i \quad \text{for all} \;\; s \in \mathscr{S} \right\}$$

Proof. The conditions in (i) are certainly necessary, and they suffice to make $\mathscr{B}$ an ideal in $\mathscr{J}$. Indeed, $\mathscr{B}$ is a linear subspace, since for any $\alpha \in \mathfrak{R}$, $b \in \mathscr{B}$, we have:

$$\alpha b = \alpha \{ebe\} = \{(\alpha e)\, be\} \in \{\mathscr{S}\mathscr{B}\mathscr{S}\} \subseteq \mathscr{B} ,$$

and since the hypothesis $sbs \in \mathscr{B}$ implies

$$2\{sbs'\} = (s + s')\, b\, (s + s') - sbs - s'\, bs'$$

also lies in $\mathscr{B}$, for all $s, s' \in \mathscr{S}$. That $\mathscr{B}$ is an ideal follows from Jacobson [1968; Lemma 1, p. 42], which shows that the operator $U_s : x \to sxs$, generates all multiplications by elements of $\mathscr{J}$, while by definition

$$s\mathscr{B}s = U_s(\mathscr{B}) \subseteq \mathscr{B} .$$

We can also give a completely elementary proof based on **Theorem 3.1(v)** above. Thus since e remains an identity element for $\mathscr{J}$, $\mathscr{B}$ will be an ideal as soon as

$$xbx \in \mathscr{B} \quad \text{for all} \;\; b \in \mathscr{B} , \quad \text{and all} \;\; x \in \mathscr{J} = \mathscr{T}_q ,$$

and therefore it is sufficient to prove

$$\{xbx'\} \in \mathscr{B} \quad \text{for all} \;\; x, x' \in \mathscr{T}_i , \;\; \text{all} \; i, \;\; \text{and} \; b \in \mathscr{B}$$

by induction on i.

The case $i = 0$ is the hypothesis of **Theorem 4.1(i)**, while for the induction step it suffices to consider only spanning elements $x = yzy$, $x' = y'z'y'$ for the space

$$\mathscr{T}_{i+1}, \text{with} \;\; y, y', z, z' \in \mathscr{T}_i .$$

Now

$$\{xbx'\} = \{yzy, \; b, \; y'z'y'\}$$

$$= 2\{y, \; z, \; \{y, \; b, \; y'z'y'\}\} + y\{z, \; y'by', \; z'\}y - 2y\{z, \; y', \; \{z', \; y', \; b\}\}y .$$

By the induction hypothesis $y'\mathscr{B}y'$, $\{z\mathscr{B}z'\}$, and $y\mathscr{B}y \in \mathscr{B}$, and also

$$\{z^*, y^*, b^*\} = 2\{z^*, \{y^*b^*e\}, e\} - \{z^*, b^*, y^*\} \in \mathcal{B},$$
$$\text{for all }\; z^*, y^* \in \mathcal{T}_i, \; b^* \in \mathcal{B}.$$

Thus we get

$$\{y, b, y'z'y'\} = 2\{\{yby'\}, z', y'\} - y'\{byz'\}y' \in \mathcal{B},$$

and with this $\{xbx'\} \in \mathcal{B}$, which completes the induction, establishing (i).

For (ii), the chain, by finite dimensionality again, becomes stationary at some

$$\mathcal{B}_t = \mathcal{B}_{t+1} = \ldots = \mathcal{B}_*.$$

Now any ideal $\mathcal{I}$ of $\mathcal{J}$ inside $\mathcal{S}$ has $\mathcal{I} \subseteq \mathcal{B}_0$, and if $\mathcal{I} \subseteq \mathcal{B}_i$ then $s\mathcal{I}s \subseteq \mathcal{I} \subseteq \mathcal{B}_i$ shows $\mathcal{I} \subseteq \mathcal{B}_{i+1}$ as well, so $\mathcal{I} \subseteq \mathcal{B}_*$. Now, $\mathcal{B}_*$ is also seen to be an ideal of $\mathcal{J}$, since $s \in \mathcal{S}$, $b \in \mathcal{B}_* = \mathcal{B}_{t+1}$ implies $sbs \in \mathcal{B}_t = \mathcal{B}_*$, and by (i) this suffices to complete the proof. ∎

4.2 Corollary.   The Jordan closure $\mathcal{J}$ of $\mathcal{S}$ can be constructed through

$$\mathcal{S} = \mathcal{U}_0 \subseteq \mathcal{U}_1 \subseteq \ldots \subseteq \mathcal{U}_\ell = \mathcal{U}_{\ell+1} = \mathcal{J}, \text{ where}$$

$$\mathcal{U}_{i+1} = \left\{ \sum t_j s_j u_j s_j \mid \text{ all } t_j \in \mathfrak{R}, \; s_j \in \mathcal{S}, \; u_j \in \mathcal{U}_i \right\}.$$

Also, $\mathcal{U}_i \subseteq \mathcal{S}_i \subseteq \mathcal{T}_i$, and $p, q \leq \ell$.

Proof.   By finite dimensionality the chain must become stationary, at $\mathcal{U}_\ell$ say, and then by **Theorem 4.1(i)**

$$s(\mathcal{U}_\ell)s \subseteq \mathcal{U}_{\ell+1} = \mathcal{U}_\ell$$

implies $\mathcal{U}_\ell$ is Jordan. ∎

We now collect and in part, restate, some elementary facts about any special $\mathcal{J}$ and subspaces of $\mathcal{S}$. Our definitions parallel the standard notions of inner and outer ideals in Jordan algebras: see Jacobson [1981; Chapter 4]. The definitions do not assume formal reality of $\mathcal{J}$ but we still assume $\frac{1}{2} \in \mathcal{J}$ and that $\mathcal{J}$ is special.

We make use of the U operator as introduced in the proof of **Theorem 4.1**:

$$U_x(a): a \rightarrow xax, \quad \text{for a given } x \in \mathcal{J}, \text{ and all } a \in \mathcal{J}.$$

For subspaces $\mathcal{A}, \mathcal{C}, \mathcal{D} \subseteq \mathcal{J}$ we'll call subspace $\mathcal{A}$

$(\mathcal{C}, \mathcal{D})$–outer if $cac = U_c(a) \in \mathcal{D}$, for all $c \in \mathcal{C}$, and $a \in \mathcal{A}$;

$(\mathcal{C}, \mathcal{D})$–inner if $aca = U_a(c) \in \mathcal{D}$, for all $c \in \mathcal{C}$, and $a \in \mathcal{A}$.

Equivalently, $\mathscr{A}$ is $(\mathscr{C}, \mathscr{D})$–outer if $\{c_1 a c_2\} \in \mathscr{D}$, for all $c_1, c_2 \in \mathscr{C}$, $a \in \mathscr{A}$, and is

$(\mathscr{C}, \mathscr{D})$–inner if $\{a_1 c a_2\} \in \mathscr{D}$, for all $c \in \mathscr{C}$, $a_1, a_2 \in \mathscr{A}$.

Moreover for $\mathscr{J}$ with identity $1_{\mathscr{J}} \in \mathscr{J}$, an inner (outer) ideal is a Jordan subalgebra $\mathscr{C} \subseteq \mathscr{J}$ that is also $(\mathscr{J}, \mathscr{C})$–inner (outer), an ideal being both inner and outer. If

$$\mathscr{C}^* \subseteq \mathscr{C}, \, 1_{\mathscr{C}} \in \mathscr{C}^*,$$

then subspace $\mathscr{C}$ is a Jordan subalgebra if it is $(\mathscr{C}^*, \mathscr{C})$–outer or inner. If $\mathscr{D}^* \subseteq \mathscr{D}$ then $(\mathscr{C}, \mathscr{D}^*)$–outer (inner) implies $(\mathscr{C}, \mathscr{D})$ outer (inner) while if $\mathscr{C}^* \supseteq \mathscr{C}$ then $(\mathscr{C}^*, \mathscr{D})$–outer (inner) implies $(\mathscr{C}, \mathscr{D})$–outer (inner). If subspaces $\mathscr{A}$ and $\mathscr{D}$ coincide then we say $\mathscr{A}$ is $\mathscr{C}$–outer (inner) if it is $(\mathscr{C}, \mathscr{A})$–outer (inner).

We can thus restate part (i) of **Theorem 4.1** as: if subspace $\mathscr{C} \subseteq \mathscr{J}$ is $\mathscr{S}$–outer then it is an ideal of $\mathscr{J}$, with $e = 1_{\mathscr{S}} = 1_{\mathscr{J}}$. Referring to the construction of the chain $\{\mathscr{B}_i\}$ in part (ii) of the theorem we see that $\mathscr{B}_{i+1}$ is the maximal subspace of $\mathscr{B}_i$ that is $(\mathscr{S}, \mathscr{B}_i)$–outer. Our next result shows that much more is true of the subspaces $\mathscr{B}_i$ and to state it we'll need some definitions: a Jordan algebra $\mathscr{J}$ is **nondegenerate** if $U_a = 0$ if and only if $a = 0$, for all $a \in \mathscr{J}$. We note that any finite dimensional $\mathscr{J}$ is nondegenerate if and only if it is semisimple, by Corollary 1 of Jacobson [1968; p. 201], so in particular $\mathscr{S} \subseteq \mathbf{S}_m$ implies all Jordan subalgebras of $\mathbf{S}_m$ are nondegenerate. An important consequence of nondegeneracy is the fact that any finite dimensional nondegenerate Jordan algebra $\mathscr{J}$ is (von Neumann) **regular**, in the sense that the range of U_a contains a, for all $a \in \mathscr{J}$: for each $a \in \mathscr{J}$, there exists an $x \in \mathscr{J}$ such that $a = axa$ (c.f. Malley [1986; p. 103]; Jacobson [1968; p. 55 and p. 179]; and especially McCrimmon [1969]).

We now demonstrate:

4.3 Theorem.   Letting $\mathscr{S} = \mathscr{B}_0$, $\mathscr{B} = \mathscr{B}_1$, and generally $\mathscr{B}_i$ be as above, we   have:

(i)   for $i > 0$, $\mathscr{B}_{i+1}$ is an $(\mathscr{S}, \mathscr{B}_{i+1})$–inner subspace of $\mathscr{B}_i$; it is the maximal such subspace if all Jordan subalgebras of $\mathscr{S}$ are nondegenerate;

(ii)   for $i \geq 0$, $\mathscr{B}_{i+1}$ is Jordan, and for $i > 0$ it is an ideal of $\mathscr{B}_i$;

(iii) $\mathscr{B}_2$ is an ideal of $\mathscr{J}$ if and only if $\mathscr{B} = \mathscr{B}_2$, if and only if $\mathscr{B} = \mathscr{B}_*$.

Proof.   Begin by recalling the definition of $\mathscr{B}_{i+1}$:

$$\mathscr{B}_{i+1} = \{b \in \mathscr{B}_i \mid sbs \in \mathscr{B}_i, \text{ for all } s \in \mathscr{S}\}.$$

Then we see that

$$2\{sbt\} = U_{s+t}b - U_s b - U_t b \in \mathscr{B}_i, \text{ for all } s, t \in \mathscr{S}, b \in \mathscr{B}_{i+1}.$$

We use this to first show that

$\mathscr{B}_{i+1}$ is $(\mathscr{S}, \mathscr{B}_{i+1})$–inner, $U_b s \in \mathscr{B}_{i+1}$ for all $s \in \mathscr{S}$ and $b \in \mathscr{B}_{i+1}$,

and hence that $\mathscr{B}_{i+1}$ is Jordan, since $b_1 . b_2 = \{b_1 e b_2\}$. Thus we need to first verify that

$$U_t U_b s = tbsbt \in \mathscr{B}_i, \text{ for all } s, t \in \mathscr{S}, b \in \mathscr{B}_{i+1}.$$

However

$$U_t U_b s = 2 \{\{tbs\}, b, t\} - \{s, b, \{tbt\}\},$$

so that

$$\{tbs\}, \{tbt\} \in \mathscr{B}_i, \text{ implies } U_t U_b s \in \mathscr{B}_i.$$

Since for $t = e \in \mathscr{S}$ we get $U_b s \in \mathscr{B}_i$, and it follows that $U_b s \in \mathscr{B}_{i+1}$ as required.

We now demonstrate the maximality of $\mathscr{B}_{i+1}$. From above we know that $\mathscr{B}_{i+1}$ is Jordan, so by the assumption on $\mathscr{S}$, $\mathscr{B}_{i+1}$ is nondegenerate and hence regular: for any $b \in \mathscr{B}_{i+1}$ there exists $x \in \mathscr{B}_{i+1}$ such that $b = bxb$. Thus

$$\mathscr{B}_{i+1} = U_{\mathscr{B}_{i+1}} \mathscr{B}_{i+1} = U_{\mathscr{B}_{i+1}} \mathscr{S}$$

To show $\mathscr{B}_{i+1}$ maximal consider now any space $\mathscr{C}$,

$$\mathscr{B}_{i+1} \subseteq \mathscr{C} \subseteq \mathscr{B}_i, \text{ that is } (\mathscr{S}, \mathscr{B}_{i+1})\text{–inner}.$$

Then $U_{\mathscr{C}} \mathscr{S} \subseteq \mathscr{B}_{i+1} \subseteq \mathscr{C}$, so $\mathscr{C}$ is necessarily Jordan and hence as before also non-degenerate, with

$$\mathscr{C} = U_{\mathscr{C}} \mathscr{C} \subseteq U_{\mathscr{C}} \mathscr{S} \subseteq \mathscr{B}_{i+1} \subseteq \mathscr{C},$$

which forces

$$\mathscr{C} = \mathscr{B}_{i+1}.$$

We've thus verified (i) and the first half of (ii), and to show that $\mathscr{B}_{i+1}$ is an ideal of $\mathscr{B}_i$, $i > 0$, consider

$$U_s (a . b) = 2\{s, a, b . s\} - (U_s a) . b,$$

for $a \in \mathscr{B}_{i+1}$, $b \in \mathscr{B}_i$. Then $U_s a \in \mathscr{B}_i$, so also $(U_s a) . b \in \mathscr{B}_i$ since $\mathscr{B}_i$ is Jordan. As $b . s \in \mathscr{B}_i$, we now get $U_{\mathscr{S}} (a . b) \subseteq \mathscr{B}_i$, so $U_e (a . b) = a . b \in \mathscr{B}_i$ and $a . b \in \mathscr{B}_{i+1}$, as required.

Finally, for (iii), if $\mathscr{B}_2 = \mathscr{B}$ then by **Theorem 4.I(i)** $\mathscr{B}$ is an ideal of $\mathscr{J}$, and $\mathscr{B} =$

$\mathscr{B}_{\bullet}$, while if $\mathscr{B}$ is an ideal of $\mathscr{J}$ then $\mathscr{B} = \mathscr{B}_2 = \ldots = \mathscr{B}_{\bullet}$. This completes the proof. ∎

3.5 Solutions of the Equation: sasbs = 0.

In the statistical context of Chapter 2 (see also Malley [1986]) the following equation arises:

for given a, b $\in \mathscr{B}$: sasbs = 0, for all s $\in \mathscr{S}$,

where $\mathscr{B}$ is the subspace of $\mathscr{S}$:

$\mathscr{B} = \{b \in \mathscr{S} \mid sbs \in \mathscr{S}, \text{ for all } s \in \mathscr{S}\}$

(= $\mathscr{B}_1$ from **Theorem 4.1(ii)**).

Here $\mathscr{S}$ is as before, that is the real linear span of a finite set of real matrices $\{m_i\}$ $\subseteq \mathbf{S_m}$. We assume as well that $\mathscr{S}$ has an identity, e $\in \mathscr{S}$, such that
es = se = e for all s $\in \mathscr{S}$,
so by Theorem 4.3, $\mathscr{B}$ is semisimple Jordan. Therefore $\mathscr{B}$ has its own identity

$1_{\mathscr{B}} \in \mathscr{B}$, with $1_{\mathscr{B}} \cdot b = b$ for all $b \in \mathscr{B}$.

Two facts are pertinent with regard to $1_{\mathscr{B}}$ and idempotents e $\in \mathscr{S}$:

(i) if e . s = s then es = se = s, for any s $\in \mathscr{S}$,
since $(e)^2 = e$ shows

ese = $2(e) . (e . s) - (e)^2 . s = 2s - s = s$, so that es = e(ese) = ese = s = e . s,

and similarly for se = s. Yet:

(ii) $1_{\mathscr{B}}$ = e the identity of $\mathscr{S}$, if and only if $\mathscr{B} = \mathscr{S}$,

since e $\in \mathscr{B}$ implies all $s^2 = ses \in \mathscr{S}$ implies $\mathscr{S}$ is already a Jordan subalgebra.
Thus a measure of care is needed with these idempotents, while we also now have the useful fact:

5.1 Lemma. For a, b $\in \mathscr{B}_{i+1}$, i ≥ 0 : $a\mathscr{B}_{i+1}b = a\mathscr{B}_i b = \ldots = a\mathscr{S}b$.
Proof. Noting that $\mathscr{B}_{i+1}$ is semisimple Jordan we know $\mathscr{B}_{i+1}$ has an identity element e., e. $\in \mathscr{B}_{i+1}$. Hence using pertinent fact (i) above it follows that for any s $\in \mathscr{S}$,
asb = (ae.)s(e.b) = a(e.se.)b,
while by **Theorem 4.3** we know that e.se. $\in \mathscr{B}_{i+1}$. Thus

$a\mathscr{S}b \subseteq a\mathscr{B}_{i+1}b,$

and the result follows since

$$\mathscr{B}_{i+1} \subseteq \mathscr{B}_i \subseteq \ldots \subseteq \mathscr{B}_0 = \mathscr{S}. \quad \blacksquare$$

In Malley [1986] it is shown that $\mathscr{B}$ is isomorphic to the space of all "optimal unbiased" estimates of estimable functions of the variance components in the mixed linear model specified by the $\{m_j\}$. Now, if it is assumed that the data vector y is normally distributed with mean zero and covariance matrix $\Sigma = s \in \mathscr{S}$, then it is known that the equation

$sasbs = 0 ,$ for all $s \in \mathscr{S},$

implies that the random quadratic forms $y'ay$, $y'by$ are statistically independent. For this reason detailed knowledge of the solutions of the equation is quite useful.

More precisely we show that the equation $sasbs = 0$, for all $s \in \mathscr{S}$, given $a, b \in \mathscr{B}$, obtains if and only if a and b have disjoint support, where the

support of $a = \oplus a_i, \; a_i \in \mathscr{B}_i,$

for $\mathscr{B} = \oplus \mathscr{B}_i$ the decomposition of $\mathscr{B}$ into simple ideals, is the set of i for which $a_i \neq 0$ (with support of 0 defined to be $\varnothing$).

Hence the message for the statistician is that the simple components $\mathscr{B}_i$ contain no non-trivial elements $a, b \in \mathscr{B}_i$ whose associated quadratic forms $y'ay$, $y'by$ are statistically independent for all covariances $\Sigma \in \mathscr{S}$, while elements b_i, b_i, with $b_i \in \mathscr{B}_i, \; b_j \in \mathscr{B}_j$, generate independent forms. Thus the Jordan algebra decompostion of $\mathscr{B}$ into simple ideals constitutes in this sense a **maximal independence decomposition** for random quadratic forms based on normal data y having **linear covariance structure:** for some set $\{m_j\}$ we have $\mathscr{S}$ exactly the span of all allowable Σ. The main result is:

5.2 Theorem.   Let $\mathscr{B} = \{b \in \mathscr{S} \mid sbs \in \mathscr{S}$ for all $s \in \mathscr{S}\}$ for $\mathscr{S} \subseteq \mathbf{S}_m$ a vector space of symmetric matrices having an identity element $e \in \mathscr{S}$, $es = se = s$ for all $s \in \mathscr{S}$. Then for given $a, b \in \mathscr{B}$ the following are equivalent:
   (i) $sasbs = 0$ for all $s \in \mathscr{S}$;
   (ii) $a\mathscr{S}b = 0$;
   (iii) $a\mathscr{B}b = 0$;
   (iv) a and b have disjoint support.
Proof.   That (ii) $\Rightarrow$ (i) is immediate. Hence we begin by showing directly that a linearization gives (i) $\Rightarrow$ (ii) and this implies equivalence of the first two statements.

Indeed, setting $s = e$ in $sasbs = 0$ yields $ab = 0$. Then replacing s by

$$\delta s + e, \quad \text{for} \quad s\delta = \pm 1$$

implies

$$0 = (\delta s + e)a(\delta s + e)b(\delta s + e) - (\delta^3)sasbs$$

$$= \delta\,[(\delta s + e)asb(\delta s + e) - (\delta^2)sasbs]$$

$$= \delta\,[\delta\{s,\ asb,\ e\} + easbe]$$

$$= \delta\,[\delta\{s,\ asb,\ e\} + asb]$$

so

$$asb = -\delta\,\{s,\ asb,\ e\} \quad \text{for} \quad \delta = \pm 1,$$

therefore $asb = 0$. Thus (i) implies (ii).

That (ii) $\Rightarrow$ (iii) is clear, since $\mathscr{B} \subseteq \mathscr{S}$, while (iii) $\Rightarrow$ (ii) follows from **Lemma 5.1**. To show (iii) $\Rightarrow$ (iv) we use:

(5.3) $\quad a\mathscr{B}c = 0 \quad$ for $a, c \in \mathscr{B} \Rightarrow a\mathscr{B}\mathscr{S}c = 0.$

This obtains by using **Lemma 1.6** for any $s \in \mathscr{S}$, $d \in \mathscr{B}$:

$$(adsc)(adsc)' = adsc^2\,sda$$

$$= a\{dsc\}csda - acsdcsda$$

$$\in (a\mathscr{B}c)sda - (ac)sdcsda = 0,$$

since $\{dsc\} \in \mathscr{B}$ by **Theorem 4.3**, and since $ac = a1_{\mathscr{B}}c = 0$. Hence $a\mathscr{B}\mathscr{S}c = 0$.

To verify (iii) $\Rightarrow$ (iv) consider now the **right $\mathscr{B}$-annihilator** of a:

(5.4) $\quad \text{Ann}_r(a) = \{c \in \mathscr{B} \mid a\mathscr{B}c = 0\}.$

This is a Jordan ideal of $\mathscr{B}$ since it is a subspace with $x \cdot c \in \text{Ann}_r(a)$ for $c \in \text{Ann}_r(a)$, $x \in \mathscr{B}$, because

$$a\mathscr{B}(xc + cx) \in a\mathscr{B}\mathscr{B}c + a\mathscr{B}c\mathscr{B} = 0,$$

by Equation 5.3, since

$$a\mathscr{B}\mathscr{B}c \subseteq a\mathscr{B}\mathscr{S}c = 0.$$

Continuing with **(iii)** $\Rightarrow$ **(iv)**, we next make use of a reduction to the components of $a = \oplus a_i$, $b = \oplus b_i$, for a_i, $b_i \in \mathscr{B}_i$:

(5.5) $asb = 0$ for all $s \in \mathscr{S}$ $\Leftrightarrow$ $a_i s b_j = 0$ for all $s \in \mathscr{S}$, all i, j.

Here the direction $(\Leftarrow)$ is clear, while the direction $(\Rightarrow)$ follows from

$$0 = e_i(asb)e_j = a_i s b_j,$$

because for the units $e_i \in \mathscr{B}_i$ we have

(5.6)   **(i)** $e_i b = b e_i = b$ $(b \in \mathscr{B}_i)$;
            (ii) $e_i b = b e_i = 0$ $(b \in \mathscr{B}_j, \; i \neq j)$; hence $\mathscr{B}_i \mathscr{B}_j = 0$, $i \neq j$.

Part **(i)** follows from an argument similar to that of fact **(i)** preceding **Theorem 5.2** above. Part **(ii)** is the fact (**Lemma 3.2**, Chapter 2) that special Jordan orthogonal ideals with units e_i are also associative orthogonal:

$$e_j e_i e_j = e_j(e_i \cdot e_j)e_j = 0 \quad \text{implies} \quad e_i e_j = 2(e_i \cdot e_j) - e_j e_i e_j = 0,$$

so

$$\mathscr{B}_i \mathscr{B}_j = \mathscr{B}_i e_i e_j \mathscr{B}_j = 0, \quad \text{for } i \neq j.$$

Thus if $asb = 0$ for all s then we have all $a_i s b_j = 0$, by a reduction through Equation 5.5 above, and $b_j \in \text{Ann}_r(a_i)$. If i is in the support of a, and j in the support of b, then $b_j \neq 0$, and $b_j \in \text{Ann}_r(a_i)$, implies $\mathscr{B}_j \subseteq \text{Ann}_r(a_i)$ by the simplicity of $\mathscr{B}_j$, and $a_i \neq 0$ implies

$$a_i(1_{\mathscr{B}})a_i = (a_i)^2 \neq 0$$

by **Lemma 1.7**. Hence a_i is not $\in \text{Ann}_r(a_i)$, which means a_i is not $\in \mathscr{B}_j$, and necessarily $i \neq j$. Thus the supports of a and b are disjoint and **(iii)** $\Rightarrow$ **(iv)**.

Finally, we verify that **(iv)** $\Rightarrow$ **(iii)**. Here, take any

$$c \in \mathscr{B}, \quad c = \oplus c_k, \quad c_k \in \mathscr{B}_k.$$

Then for $a = \oplus a_i$, $a_i \in \mathscr{B}_i$, $b = \oplus b_j$, $b_j \in \mathscr{B}_j$, and from $a_i c_k = 0$ for $i \neq k$, and using Equation 5.6(ii) we see

$$acb = \Sigma\, a_i c_k b_j = \Sigma\, a_i c_i b_j = 0,$$

as the supports of a and b are disjoint, and this completes the proof. ∎

Applying mtehods similar to those above finally shows:

5.7 Theorem For any given elements a, b $\in \mathscr{J} = \mathscr{J}(\mathscr{S})$ the following are equivalent:

    (i)   sasbs = 0 for all $s \in \mathscr{J}$;

    (ii)   sasbs = 0 for all $s \in \mathscr{S}$;

    (iii) a$\mathscr{J}$b = 0 for all $s \in \mathscr{J}$;

    (iv) a$\mathscr{S}$b = 0 for all $s \in \mathscr{S}$;

    (v)   a and b have disjoint support in $\mathscr{J} = \oplus\mathscr{J}_i$.

References for Chapter 3

Braun, H. and Koecher, M. [1966]. **Jorden-algebren**. Grundlehren der Mathematischen Wissenschaften, 128. Springer–Verlag, Berlin.

Emch, G. [1972]. **Algebraic Methods in Statistical Mechanics and Quantum Field Theory**. Wiley–Interscience, New York.

Greub, W. [1978]. **Multilinear Algebra**. 2nd Edition. Springer–Verlag, New York.

Herstein, I. [1968]. **Noncommutative Rings.** Carus Mathematical Monograph No. 15, Mathematical Association of America, Wiley, New York.

Jacobson, N. [1968]. **Structure and Representations of Jordan algebras**. American Mathematical Society, Colloquium Publication Volume XXXIX, Providence, Rhode Island.

Jacobson, N. [1981]. **Structure Theory of Jordan Algebras**. University of Arkansas Lectures Notes in Mathematics, Vol. 5, Fayetteville, AK.

Jensen, S. T. [1988]. "Covariance hypotheses which are linear in both the covariance and the inverse covariance." **Annals of Statistics, (16)**, 302–322.

Jordan, P., von Neumann, J., and Wigner, E. [1934]. "On an algebraic generalization of the quantum mechanical formalism." **Annals of Math., (35)**, 29–64.

Malley, J. [1986]. **Optimal Unbiased Estimation of Variance Components**. *Lecture Notes in Statistics*, Vol. 39. Springer–Verlag, New York.

McCrimmon, K. [1969]. "Nondegenerate Jordan rings are von Neumann regular." **Journal of Algebra, (11)**, 111–115.

Schafer, R. [1966]. **An Introduction to Non-Associative Algebras.** Academic Press, New York.

Chapter 4   Jordan Algebras, the EM Algorithm, and Covariance Matrices

4.1. Introduction.

Using Jordan algebras, Galois field theory and the EM algorithm we show how to obtain either essentially closed-form or simplified solutions to the ML equations for estimation of the covariance matrix for multivariate normal data in the following situations:

(a)   Data having a patterned covariance matrix; equivalently, data with a linear covariance structure. This case includes the problems of variance and variance-covariance components estimation; unbalanced repeated measures designs; and some time series models;

(b)   Data vectors with values missing completely at random; in particular, retrospective analyses of long term, clinical trials, incomplete repeated measures, and time series;

(c)   The intersection of (a) and (b): multivariate normal data assumed to have a linear covariance structure but with some values missing (completely at random);

In all these cases our method generates an iterative scheme that has guaranteed convergence to the unique ML solution, whenever it exists. The method is often in "nearly" closed–form, since the M–step is shown to effectively vanish: by using a proper choice of basis matrices (that is, by reparametrizing the problem) the current estimate of the parameter vector simply re-enters the E-step for the next update of the complete data vector, as the M–step is then reduced to multiplication by the identity matrix. In all cases, the E-step is relatively immediate, being formed from simple matrix multplications and a single submatrix inversion.

After presenting the general solution to the patterned (= linearly constrained) covariance matrix estimation problem we consider some special cases.

In particular, certain constrained covariance matrices have interesting Jordan algebra related solutions that are not of the type treated in the first main theorem. One interesting type considered is that of covariance matrices with a constant (unknown) diagonal. This investigation leads to commuting Jordan algebras and the algorithm yields an M–step based on a set of commuting, symmetric matrices. Since such matrices are simultaneously diagonalizable, and it is known that the likelihood equations then considerably simplify, though still requiring an iterative mechanism for solution.

A novel, global feature of the methods proposed here is the introduction in some cases of additional parameters, ones not included in, or functions of, the original parameter vector. In a concluding section we consider in what senses the added parameters are nuisance parameters, or whether they contain important information. Using the information matrix based on different combinations of observed data, complete data, and original parameters, added parameters, two views on this are developed, leading to opposite conclusions. This discussion requires rather diverse methodologies: the differential geometric approach to statistical estimation and information, as described for example by Amari [1985, 1987]; and also one of the new results, derived earlier in Chapter 2, on the independence of random quadratic forms in normal variables.

Finally, to make connection with more recent research, we alert the reader to the now sizable literature on the Gibbs Sampler. and related methods of "data augmentation." Some of the basic references for the Gibbs Sampler can be found in Wilson, Malley, et al. [1992], where it is shown how the Sampler can be quite effectively implemented on an Intel iPSC/860, massively parallel computer. For more accesses to this literature see Gelfand and Smith [1990].

4.2 The General Patterned Covariance Estimation Problem

We begin with an overview and some background.

Maximum likelihood estimation of the covariance matrix for multivariate normal data, when the matrix has a known linear patterned structure, has been extensively studied: see for example Anderson [1969], Szatrowski [1980], Rubin and Szatrowski [1982].

In principle the EM algorithm can be used, but it is often very slow to converge, especially when either or both of its two parts (the E–Step, and the M–step) are computationally difficult. In certain cases, the pattern of the covariance matrix permits a closed form, non-iterative solution to the maximum likelihood (ML) equations. However, these special cases are probably only rarely met in practice.

We present a nearly closed form solution to the general patterned covar-iance matrix ML estimation problem, where "nearly" means that the M–step of the algorithm is trivial: it is a multiplication by a fixed, known, matrix that is easily calculated and is independent of the data.

Our method applies to all linearly constrained covariance matrices (e.g. Toeplitz). It has a natural extension to covariance estimation in the presence of data missing completely at random. The method also yields a simplification of the EM procedure for nonlinearly constrained covariance matrices, which also remains valid in the presence of missing data.

The procedure is illustrated for a covariance matrix having Toeplitz form, a pattern type that occurs, for example, in all stationary time series data.

We turn now to a more detailed introduction.

A problem of considerable practical importance is the optimal estimation of the covariance matrix for multivariate normal data, when the data is known to have **linear covariance structure**, that is: the covariance matrix is expressible as a real linear combination of known symmetric matrices. (Evidently Anderson [1969] first introduced the term "linear covariance structure," while other writers use the equivalent term "patterned" covariance matrix for this same idea; we'll use the terms interchangeably.)

Recent work by Dembo et al. [1989] on maximum likelihood (ML) estimation of covariance matrices of Toeplitz form (which arise in stationary time series data) speaks well to both the practical significance and the theoretical challenge of the problem. See also: Srivastava [1966], Burg et al. [1982], Fuhrmann and Miller [1988], Dembo [1988], and Lev-Ari et al. [1989]. Linearly structured covariance matrices also arise in repeated measures designs, growth models, and, finally, a particular case of the problem is precisely that of ML estimation of variance components.

Earlier work by Anderson [1969], Rogers and Young [1977], Szatrowski [1980], and Rubin and Szatrowski [1982] examined certain spaces of linearly structured covariance matrices (technically: Jordan algebras) that lead to simplified forms for the ML equations. Unfortunately, the required Jordan algebra structure needed for this approach applies in only a handful of practical cases, and, in particular, does not directly apply to covariance matrices with an arbitrary, non-banded form.

Interestingly, the method given in Dembo et al. [1989] for Toeplitz matrices, while using the EM algorithm, implicitly does employ a Jordan algebra approach, but this fact was not noted there. Also, the REML method of variance component estimation can be written as an implementation of the EM algorithm (see Dempster, Laird and Rubin [1977], or Little and Rubin [1987; p. 149–152]) and has a trivial M–step because an elementary Jordan algebra (a space of strictly diagonal covariance matrices) is involved in the EM formulation. Further details about both of these comments appear below.

In principle the EM algorithm is applicable to all forms of the patterned covariance matrix problem, but, as is the case with many other invocations of the method, the algorithm can be very slow to converge, especially when the individual iterations are non-trivial.

Here we show how to enlarge the original matrix problem to an M L estimation problem for which the Jordan algebra condition is always satisfied. The linear constraints of the original problem are then imposed via the wholly standard methhod of Lagrange undetermined multipliers. It is on this higher dimensional, constrained model that the EM algorithm is applied. The resulting equations for the M–step are always linear and always have a unique, explicit solution (when no data is missing).

The extension of the method to problems with missing data begins with a different Jordan algebra result, after which the first Jordan algebra technique is applied.

Remarks.

(i)   Magnus [1988] presents an elegant, unified study of a wide variety of results related to patterned matrices and "linear structures," as he refers to them. In particular he obtains the mean and covariance for the asymptotic multivariate normal distribution for the general, linearly constrained ML estimation problem; see Magnus [1988; §10.7]. This result is readily applied to the point estimates we obtain here.

On the other hand, the problem of nonlinear constrained ML estimation (still in the context of the EM algorithm) is considered in Synder and Miller [1991; see §3.4]. In their treatment important, but quite different, constraints are examined, for example: limited bandwidth estimation.

(ii)   Left unresolved here is the problem of multiple local (or equal global) maxima that may arise as solutions to the M–step in the presence of missing data. When non-unique ML solutions, or multiple, equal global solutions are known or suspected, our general recommendation is application of the Gibbs Sampler or one of the many, recently developed "data augmentation" methods; see for example Gelfand and Smith [1990], or as mentioned above, Wilson and Malley, et al. [1992].

(iii)   Also left unresolved here is the perennial problem of obtaining a positive definite solution. As the general condition for positive definiteness of a matrix is a nonlinear inequality condition, our method cannot be obviously extended to resolve this issue. In the particular case of positive semidefinite variance component estimation, however, there is a simple, quadratic reparameterization and a set of side conditions that can be used to guarantee convergence to a positive definite estimate of all the components, given that a unique ML estimate exists. This leads to solving a simultaneous set of quadratic equations at the M–step; details won't be pursued here.

(iv)   Finally, although the algorithm is simplified by our method, we do not address the actual speed of convergence. As often observed in practice, the E M algorithm may have elementary E– or M–steps but may still be slow in its approach to the maximum.   □

A comment about notation in this chapter: we continue our practice of adopting notation native to the technical subject matter at hand. Since, for example, our results here are not primarily ring–theoretic, and since the statistical literature on the EM algorithm is fairly consistent notationally, we'll denote random variables by X, Y, Z, etc., and write matrices using non-bold capitals appearing earlier in the alphabet: A, B, C, etc. Other changes will be transparent, and we hope, uneventful.

4.3. Precise Statement of the Problem.

Let $\mathbf{S}_m$ denote the real vector space of all order r, real symmetric matrices, and write **Skew-S**$_m$ for the space of order r, real skew–symmetric matrices; $\mathbf{I}_r$ will denote the order r identity matrix.

Assume the r.v. X is r-dimensional, multivariate normal, mean zero, with a linear covariance structure:

X is distributed as $\mathcal{N}_r(0, \Sigma_X)$,

with

$$\Sigma_X = \sum \zeta_i G_i, \quad \text{for } 1 \le i \le k,$$

where

$$\zeta = (\zeta_1, \dots, \zeta_r)^T \in \mathfrak{R}^r,$$

is a real r-vector, and the G_i are all known elements of $\mathbf{S}_m$.

Our task is to estimate ζ using maximum likelihood; it is assumed that the true, but unknown covariance matrix Σ_X is any positive definite element in the linear space in $\mathbf{S}_m$ spanned by the matrices $\{G_i\}$. We use the EM algorithm (see for example Dempster et al. [1977], or Little and Rubin [1987]), and properties of Jordan algebras. We recall that a central reference for Jordan algebras is Jacobson [1968]; see also Malley [1986], and Chapters 2 and 3 above.

Some preliminary remarks are in order.

Remarks.

(1)   We need assume only a single realization of X; the extension to sample sizes greater than one is immediate. Given t realizations of X, $X = x^{(i)}$, with $1 \le i \le t$, define the matrix C in $\mathbf{S}_m$ by:

$$C = (1/t)[x^{(1)} \mid \dots \mid x^{(t)}][x^{(1)} \mid \dots \mid x^{(t)}]^T.$$

In the proposed method the data enters only through the matrix C of products and cross–products.

(2)   The extension to X with mean not zero, equal to Db say for design matrix D and design parameter b, is straightforward. Write

$$M = I_r - D(D^T D)^- D^T \quad \text{with } (D^T D)^- \text{ any g-inverse for } D^T D.$$

Then there is a standard factorization of M such that

$$M = KK^T, \quad K^T K = I_v \text{ (v = the rank of M),} \quad \text{and } K^T D = 0.$$

Then $X_* = K^T X$ has mean zero, and the matrices $\{G_i\}$ are replaced by the set

$$\{K^T G_i K\} \text{ in } \mathbf{S}_v.$$

(3)   Using Lemma 1 below it is always possible to reparameterize the problem so that the set of matrices $\{G_i\}$ can be replaced by a set whose members are all positive definite. Then, for example, the set $\{K^T G_i K\}$ will also be all positive definite, hence all non-zero. Therefore none of the original ζ_i will, after the replacement, be associated with zero matrices: none will be omitted in the transformed model, and ML estimates can be obtained for all components ζ_i. ☐

4.4. The Key Idea of Rubin and Szatrowski.

Rubin and Szatrowski [1982] (hereafter often denoted by just RS) introduced the following idea: embed the covariance matrix of X in a larger one, a covariance matrix such that the EM algorithm for finding ML estimates is more tractable. They studied embeddings such that the M–step reduces to a linear matrix equation for the new parameter vector (hence also for the original vector ζ), given the current value of the complete sufficient statistic obtained from the E–step.

That, remarkably, such a linearization is possible at all (independently of any embedding scheme) was evidently first observed by Anderson [1969], while Szatrowski [1980] expanded on the idea, making important connections with Gauss-Markov and UMVUE estimation of the covariance matrix.

Ultimately, we summon the power of Jordan algebras in two distinct ways, where:

Definition.   A real, linear space $\mathscr{S}$ of real, symmetric matrices is a **Jordan algebra** if and only if: $s^2 \in \mathscr{S}$ for every $s \in \mathscr{S}$.

Alternative definitions for a Jordan algebra (of real symmetric matrices) are given in Malley [1986], and of course in Chapter 2 above. The definition above is all that we will need here: it is completely equivalent to those in Chapter 2, since we assume the the identy matrix $\mathbf{I} \in \mathscr{S}$.

Using one of the alternative definitions, a real, linear space of symmetric matrices is also a Jordan algebra if and only if the space contains all its inverses. This is the basic, enabling fact that permits rewriting the full set of ML equations in a linear form; see Anderson [1969; p. 62, Equation (39)], and Szatrowski [1980; p. 807, proof of Theorem 4].

Example: A Toeplitz Matrix Embedding.
Assume that for real a, b, and c the covariance matrix Σ_X for X is any positive definite, order three matrix of the form:

$$\begin{bmatrix} a & b & c \\ b & a & b \\ c & b & a \end{bmatrix}.$$

This can be expressed as the sum:

$$\Sigma_X = aG_1 + bG_2 + cG_3,$$

for matrices G_1, G_2, G_3:

$$G_1 = \begin{bmatrix} 1 & 0 & 0 \\ 0 & 1 & 0 \\ 0 & 0 & 1 \end{bmatrix}, \quad G_2 = \begin{bmatrix} 0 & 1 & 0 \\ 1 & 0 & 1 \\ 0 & 1 & 0 \end{bmatrix}, \quad G_3 = \begin{bmatrix} 0 & 0 & 1 \\ 0 & 0 & 0 \\ 1 & 0 & 0 \end{bmatrix}.$$

It is readily checked that the space $\mathscr{S}$ spanned by the G_i is **not** a Jordan algebra: $(G_3)^2$ is not in $\mathscr{S}$. However, the space of all order four matrices of the form

$$\begin{bmatrix} a & b & c & d \\ b & a & b & c \\ c & b & a & b \\ d & c & b & a \end{bmatrix}$$

does form a Jordan algebra. Note that Σ_X appears identically in the upper left corner in the matrix above. Note also that existence of a set of values of the parameters a, b, and c for which Σ_X is positive definite need **not** imply that there exists **any** real value d such that the order four matrix is also positive definite. ◻

This last problem is one of the central obstacles to applying the RS method to an arbitrary covariance matrix with linear structure. However, independently of any Jordan algebra considerations, considerable work has been done on just this problem of obtaining positive definite embeddings; see for example Grone et al. [1984] which derives a necessary and sufficient (graph–theoretic but still easy to apply) condition for all such embeddings.

Another problem facing a wider application of the RS method is that it is not known in general how to find any space of larger matrices that is Jordan, given the requirement that the original space of covariance matrices occupies the upper left corner (be embeddable).

Dembo et al. [1989] were able to derive an (pure) existence proof for Jordan embeddings for every space of Toeplitz matrices, an embedding scheme such that if the original covariance is positive definite, then so is the larger matrix. Their method is, it must be emphasized, primarily an existence proof only, in that the order of the larger matrix can evidently grow without limit, depending on the range of eigenvalues for the original space of allowable covariance matrices; see also Dembo [1988].

We observe that their method nonetheless implicitly uses the Jordan algebra approach in that they propose embedding the space of Toeplitz matrices in question in a family of **circulant** matrices. These patterned matrices can be written as linear combinations of powers of a single permutation matrix; see for example Lancaster [1969; p. 267]. Therefore, the larger family forms a Jordan algebra: it contains all its squares. As noted above, and discussed below, it then follows that the M–step is linear, and this is form of the result they obtain.

In contrast, the method proposed here has the following properties:

(a)   it obtains a Jordan embedding for an arbitrary (e.g. not necessarily Toeplitz) space of covariance matrices having linear structure;

(b)   the embedding takes the linear structure of the original matrices into account with the use of side conditions in the M–step of the EM algorithm;

(c)   if the space of allowable covariances is contained in S_m then the embedding is of order exactly 2r; that is, the order of the required embedding is always precisely known; and

(d)   finally, every positive definite Σ_X occupies the upper left corner of a matrix (in a Jordan algebra of order 2r) that is again positive definite.

4.5 Outline of the Proposed Method.

Begin by adopting the strategy of *Rubin and Szatrowski sketched above. That is, given the r.v. X, adjoin a "missing," or "latent," or "virtual" (our term) normal r.v. Z such that*

$$Y = \left[\frac{X}{Z}\right] \in \mathfrak{R}^{2r}$$

has covariance matrix Σ_Y, where $\Sigma_Z = \Sigma_X$, and where the real linear space generated by all allowable covariance matrices for Y is assumed to form a Jordan algebra. To this enlarged estimation problem we apply the EM procedure.

Thus, given data for X, let the matrix of cross products be denoted by C, as defined earlier [Remark (1), in §2], and partition Σ_Y as:

$$\Sigma_Y = \begin{bmatrix} \Omega_{11} & \Omega_{12} \\ \Omega_{21} & \Omega_{22} \end{bmatrix} = \Omega,$$

where

$$\Sigma_X = \Omega_{11} = \Omega_{22} = \Sigma_Z, \quad \text{and} \quad \Omega_{12} = (\Omega_{21})^T .$$

As displayed in Rubin and Szatrowski [1982], the **E–step**, given the data matrix C and the i-th current value for the matrices in the partitioning of Ω, is the conditional expectation of the complete sufficient statistic C_Y

$$C_Y = (1/t)[y^{(1)} | \ldots | y^{(t)}][y^{(1)} | \ldots | y^{(t)}]^T$$

is given by:

$$E(C_Y | C, \Omega^{(i)}) = \begin{bmatrix} \mathscr{A} & \mathscr{B} \\ \mathscr{B}^T & \mathscr{C} \end{bmatrix},$$

where $\Omega^{(i)}$ is the i-th current value for Ω and:

$$\mathscr{A} = C, \qquad \mathscr{B} = C(\Omega_{11}^{(i)})^{-1}\Omega_{12}^{(i)},$$

$$\mathscr{C} = (\Omega_{22}^{(i)})^{-1} - \Omega_{21}^{(i)}[(\Omega_{11}^{(i)})^{-1} - (\Omega_{11}^{(i)})^{-1}C(\Omega_{11}^{(i)})^{-1}]\Omega_{12}^{(i)}.$$

After presenting two results, the **M–step** for our implementation of the EM algorithm will be described. For future use, let:

$$E(C_Y | C, \Omega^{(i)}) = \mathscr{E}^{(i)}.$$

Remark. In the shift from the r.v. X to the r.v. Y, we have potentially enlarged the parameter space, and the ML estimates for the parameterization of Σ_X are found as a subvector of the parameterization of Σ_Y. It is evidently not universal that a subvector of an ML vector estimate is again an ML estimate for any meaningful or related submodel. More specifically for our problem, it is not obvious that the ML estimate for Σ_X can be recovered uniquely from the ML estimate for Σ_Y. That this is indeed the case for the method we propose follows from the fact that the log likelihood for Y decomposes nicely into a sum of functions, one term of which is the log likelihood for X.

To rigorously demonstrate this, consider the technique discussed in Little and Rubin [1987, Chapter 6] of linearly decomposing the log likelihood for the problem. We revert here to the conventional notation for covariance matrices: see Anderson [1984, §2.5].

As above, writing $Y = (X, Z)$ we find that the log likelihood, $\mathscr{L}$, for Y can then be expressed as:

$$\mathscr{L}(\Sigma_{11}, \Sigma_{12}, \Sigma_{22} \mid X, Z) = \mathscr{L}_*(\Sigma_{11} \mid X) + \mathscr{L}_{**}(\psi, \Sigma_{11.2} \mid X, Z),$$

where

$$\Sigma_{11} = \Omega_{11}, \quad \Sigma_{12} = \Omega_{12}, \quad \Sigma_{22} = \Omega_{22},$$

and $\mathscr{L}_*$ is the log of the likelihood for Σ_{11}, given X, and $\mathscr{L}_{**}$ is the log of the conditional likelihood, given X and Z, for

$(\psi, \Sigma_{11.2})$,

where

$\psi = \Sigma_{12}\,\Sigma_{11}^{-1}$
is the matrix of regression coefficients, and where

$\Sigma_{11.2} = \Sigma_{22} - \Sigma_{12}\,\Sigma_{11}^{-1}$ is the conditional variance.

We note that given any fixed, allowable values of Σ_{11}, $\Sigma_{11.2}$, and ψ, we can explicitly and bijectively solve for Σ_{12} and Σ_{22}. Hence using the invariance of ML estimation under a bijection it follows that to find the maximum of the log likelihood $\mathscr{L}$ above we need only find maxima of the separate terms in the sum, that is the ML estimate of

Σ_{11},

and then also of

$(\psi, \Sigma_{11.2})$,

where Σ_{11} is now fixed in the second maximization. (Note that all maxima are taken subject to whatever, not necessarily linear, constraints are present for Σ_{11}, Σ_{12}, and Σ_{12}.) In other words, because of bijectivity, fixing Σ_{11} in the second maximization still allows $(\psi, \Sigma_{11.2})$ to assume the value that, when solved for

$(\Sigma_{11},\ \Sigma_{12},\ \Sigma_{22})$

gives exactly the required maximization for $\mathscr{L}$.

This completes our proof, and also slightly generalizes the discussion given in Little and Rubin [1987], in that we allow for new parameters in Ω. $\square$

4.6 Preliminary Results.

Two lemmas are needed: the first is standard, the second perhaps less so.

Lemma 1. Given A in $\mathbf{S}_m$ there always exists a real $\delta > 0$ such that the translate $I_r + \delta A$ is positive definite.

Proof. (Sketch of two proofs)
(1)   From Rao [1973; p. 62]:

$$\lambda_{max} = \max\left(\frac{u^T A\,u}{u^T u}\right) \quad \text{for all } u \in \mathfrak{R}^r,\ u \neq 0,$$

where λ_{max} is the largest eigenvalue of A. Choose any δ such that

$$\delta > 1/|\lambda_{max}|,$$

or any real δ greater than zero if λ_{max} is zero. Then:

$$u^T(I_r + \delta A)u > 0, \quad \text{for all } u \neq 0,$$

and positive definiteness of the translate $I_r + \delta A$ follows. ■ **(end of Proof 1)**

(2)   In the Zariski topology for S_m the connected component of the identity I_r consists of exactly the convex space of positive definite elements of Sym(r); see, for example, Jacobson [1968; p. 55]. Choosing any δ sufficiently small (positive or negative) will move the translate $I_r + \delta A$ into this component. ■ **(end of Proof 2)**

Remarks.
We now verify Remark **(3)** made in **§4.3**, make some additional comments, and list a series of steps that reduces the original problem to a more canonical form. This is done by pulling together or extending the accumulated technical details already given:
(1)   the original $\{G_i\}$ can be replaced by positive definite matrices that are known, fixed translates of the identity, if the linear span of the original set $\{G_i\}$ contains the identity;
(2)   in the case that the span of the $\{G_i\}$ does not contain the identity, a similarity transformation returns positive definite matrices, again with no loss in generality for ML estimation; that is, we can henceforth assume that the span of the $\{G_i\}$ contains the identity matrix;
(3)   a translation of the basis set $\{G_i\}$ (to gain positive definiteness) will change the form of the linear constraints, but in a fixed, known way that is easily calculated, yielding a new set of linear constraints;
(4)   a transformation using the matrix K above in Remark **(3)** (to get a r.v. X with mean zero) also will result in a new set of linear constraints, but again in a fixed, known way that is readily obtained, and, again, with no loss in generality for ML estimation;
(5)   the comments above do not require that the space $\mathscr{S}$ spanned by the set $\{G_i\}$ forms a Jordan algebra. On the other hand, if $\mathscr{S}$ is Jordan then the space spanned by the positive definite replacements is Jordan, as also is the space spanned by the replacements obtained through a similarity transformation. ❑

Lemma 2. Suppose a given B is in Sym(r), while F is any element of **Skew-S$_m$** .
Then there exists a real constant $\delta > 0$ such that the matrix Ω is positive definite, where:

$$\Omega \; = \; \begin{bmatrix} B & \delta F \\ -\delta F & B \end{bmatrix}$$

Proof. This results appears, for example, in Eaton [1983; pp 368–370], as part of a discussion of group invariant covariance matrices. A sketch of the proof follows.

Let u_i, $i = 1, 2, 3$, be the standard orthonormal basis for $\Re^3$, and write:

$$P \; = \; u_1 u_1^T , \qquad Q \; = \; I_r \; - \; P \; = \; u_2 u_2^T \; + \; u_3 u_3^T ,$$

and for

$$v_2 \; = \; (1/\sqrt{2})(u_2 \; + \; u_3), \qquad v_3 \; = \; (1/\sqrt{2})(u_2 \; - \; u_3) ,$$

let

$$J \; = \; v_2 v_3^T \; - \; v_3 v_2^T .$$

Note that P and Q are in $\mathbf{S_3}$, while J is in **Skew–$\mathbf{S_3}$**.

Now, for any positive definite matrices A and B in $\mathbf{S}$, one checks that
$$W \; = \; P \otimes A \; + \; Q \otimes B$$
is a positive definite element of $\mathbf{S_3}$. From a standard extension of Lemma 1, there exists a real $\delta > 0$ such that:
$$W_0 \; = \; W \; + \; \delta(J \otimes F) \; = \; W \; + \; J \otimes \delta F$$
is again positive definite.

Next, let $\mathscr{L}$ denote the space of real matrices of order 3r. Let $\{w_j\}$ be the standard orthonormal basis for $\Re^r$, and consider the set
$$\{u_i \otimes w_j\} \qquad \text{for} \quad 1 \le i \le 3, \; 1 \le j \le r.$$
The set forms an orthonormal basis for the space of all linear transformations of $\mathscr{L}$ onto itself, and this permits a matrix representation for W of the form:

$$W \; = \; \begin{bmatrix} A & 0 & 0 \\ 0 & B & \delta F \\ 0 & -\delta F & B \end{bmatrix} .$$

With A positive definite it follows that Ω defined as above is itself positive definite, and the proof is complete. ■

4.7 Further Details of the Proposed Method.

Begin by considering a special case of the matrix Ω just defined, that is:

$$\Omega \; = \; \begin{bmatrix} \Sigma_X & \delta F \\ -\delta F & \Sigma_X \end{bmatrix} ,$$

where Σ_X is any positive definite element of the space of allowed, linearly constrained covariance matrices for the r.v. X, and F is any element of **Skew–S$_r$**. As just shown for any given Σ_X there exists a matrix δF such that Ω is positive definite.

Next, consider ML estimation for a r.v. Y of dimension 2r that is multivariate normal, mean zero, and that has a positive definite covariance matrix of the form Ω just given. Note that the space of allowable covariance matrices (spanned by all the allowed linearly constrained Σ_X and all F in **Skew–S$_r$**) is not necessarily Jordan: this would entail some algebraic intertwining between allowable Σ_X and matrices F, but, for any given space of allowable covariance matrices such a connection may not be obtainable for a given Σ_X in the space and any matrix F in **Skew–S$_r$**.

This last problem is resolved via Lagrange undetermined multipliers in the M–step of the EM algorithm. To this end, consider the unconstrained estimation problem; the linear constraints are then insinuated into the M–step in the usual way.

Therefore, begin now with matrices Ω of the form:

$$\Omega = \begin{bmatrix} B & F \\ -F & B \end{bmatrix},$$

for B an arbitrary element of **S$_r$**, and F is an arbitrary element of **Skew–S$_r$**. Matrices with this type form a real vector space: they have a linear pattern. One convenient basis consists of a set formed by the union of two sets of matrices, $\{U_{ij}\}$ and $\{U_{kl}\}$; using $\{e_{ij}\}$ to denote the unit matrices of order r (where e_{ij} has a one at the ij-th position and is zero elsewhere) define:

$$U_{ij} = \begin{bmatrix} +1 & 0 \\ 0 & +1 \end{bmatrix} \otimes \{e_{ij} + e_{ji}\} \qquad 1 \le i, j \le r;$$

$$U_{kl} = \begin{bmatrix} 0 & +1 \\ -1 & 0 \end{bmatrix} \otimes \{e_{kl} - e_{lk}\} \qquad 1 \le k, l \le r.$$

Let's renumber the set $\{\{U_{ij}\}, \{U_{kl}\}\}$ more simply as $\{U_s\}$, where $1 \le s \le q$, with $q = r^2$. Then the space of matrices Ω is:

$$\Omega = \sum \eta_s U_s , \qquad \text{for } \eta = (\eta_1, \dots, \eta_q)^T, \text{ say.}$$

In this representation the vector η is unconstrained, and the space forms a Jordan algebra: it contains the squares of all its elements (checked most easily using the (B, F) matrix form above for Ω).

Remark. In choosing a basis for Ω it is essential to take note of a problem and modify the basis set $\{U_s\}$ slightly. The problem is this: the parameter space for

ML estimation in the exponential family (for the r.v. Y) must have a non-empty (topological) interior, in order to assure the existence and uniqueness of an M L estimate. The natural parameter space for Y is convex (positive sums of positive definite matrices are again positive definite), however, the parameter space must also contain a non-empty open set in q space, such as a q–dimensional rectangle. For the mathematical details, see for example, Bickel and Doksum [1977; Theorem 3.3.2, p. 106], or Brown [1986; Chapter 5].

More concretely, using the matrices U_s above gives rise to a subspace of matrices that happens to contain **no** allowable positive definite covariance matrix: the space spanned by just the set $\{U_{kl}\}$ is such a subspace. Thus using the $\{U_s\}$ exactly as above leads to an implicit restriction on the vector η and the parameter space, namely to a region that excludes the space spanned by just the set $\{U_{kl}\}$, and so the parameter space may not obviously contain any q–dimensional open set. The problem is however readily resolved: examining the eigenvalues for the matrices $\{U_s\}$ shows that simply adding $2I_{2r}$ to each U_s returns a set of positive definite matrices. Then, the vector parameter space contains (at least) the open rectangle of all points η such that each component η_i is strictly positive.   □

Continuing with the detailed description of our procedure, we follow in the footsteps of Anderson [1969], Szatrowski [1980], and Rubin and Szatrowski [1982], and find that the **M–step** for the unconstrained maximization is linear.

Let

$$\mathcal{U} = \text{tr}[U_s U_t], \quad \text{for } 1 \le s, t \le q.$$

Then the M–step is given by:

$$\mathcal{U}\eta^{(i+1)} - \tau^{(i)} = 0,$$

where

$$\eta^{(i+1)} = \text{the updated value for } \eta, \text{ given } \eta^{(i)},$$

and

$$\tau^{(i)} = [\text{tr}(U_s \mathscr{E}^{(i)})], \quad 1 \le s \le q.$$

(Recall the definition of $\mathscr{E}^{(i)}$ given at the end of §4.3)

Remark. The REML method for variance component estimation can be represented as an instance of unconstrained EM estimation of a covariance matrix: see Dempster, Laird and Rubin [1977; p. 17–18]. The algorithm is applied to "missing" data having a strictly diagonal covariance matrix, $\mathscr{D}$, of the form

$$\mathcal{D} = \text{diag}(\sigma_1^2 I_{c_1}, \ldots, \sigma_m^2 I_{c_m}).$$

Since the space spanned by all matrices of this form (where the variance components are allowed to be negative) are a Jordan algebra, it follows that the M–step
of the method is linear, in fact, simply multiplication by a fixed constant; see their
Equation 4.4.8. Once again, as with the method of Dembo et al. [1989], an estab-
lished EM solution for obtaining ML estimates is linearized at a key step because
of tacit use of a Jordan algebra. ❑

Consider now the known, linear constraints on allowable Σ_X. For some matrix
H let these the conditions be expressed as the set of linear equations:
    $H\eta = 0$.
Note that since the matrices F in **Skew–S**$_r$ are arbitrary, the matrix H only spec-
ifies constraints on elements of the matrix B appearing in Ω. Also, with no loss in
generality H is taken to have full rank, so the standard, non-zero Jacobian condition on systems of constraints for the Lagrangian method is met.
    The complete M–step for obtaining $\eta^{(i+1)}$, given the constrains on B is the pair
of equations:

(a) $\mathcal{U}\eta^{(i+1)} - \tau^{(i)} - H^T\lambda^{(i+1)} = 0;$

(b) $H\eta^{(i+1)} = 0.$

Here $\lambda^{(i+1)}$, in $\Re^p$ (p = rank H), is the vector of multipliers for the (i+1)-th M–step.
    Since the basis set $\{U_s\}$ is linearly independent, one can show that $\mathcal{U}$ is positive
definite, hence invertible. Equation (a) is therefore equivalent to (a*):

(a*) $\eta^{(i+1)} - \mathcal{U}^{-1}\tau^{(i)} - \mathcal{U}^{-1}H^T\lambda^{(i+1)} = 0.$

Multiplying (a*) by H and using (b) gives the solution for $\lambda^{(i+1)}$:

$$\lambda^{(i+1)} = -(H\mathcal{U}^{-1}H^T)^{-1}H\mathcal{U}^{-1}\tau^{(i)},$$

and, therefore, the solution for $\eta^{(i+1)}$ is:

$$\eta^{(i+1)} = [\mathcal{U}^{-1} - \mathcal{U}^{-1}H^T(H\mathcal{U}^{-1}H^T)^{-1}H\mathcal{U}^{-1}]\tau^{(i)}.$$

As promised, the constrained M–step has been linearized: it is multiplication by
a fixed, known matrix that is data-independent. Finally, letting

$$\Omega^{(i+1)} = \sum \eta_s^{(i+1)} U_s,$$

and using this update in the equations for $\mathcal{A}$, $\mathcal{B}$, and $\mathcal{C}$ yields an input for the next E–step, from which values for $\mathcal{E}^{(i+1)}$ and $\tau^{(i+1)}$ are obtained.

This completes the presentation of our proposed method.

Summarizing: it always leads to a linear equation at the M–step, it requires application of the Rubin and Szatrowski procedure for matrices of order exactly 2r, and it always provides a positive definite embedding of the original, allowable covariance matrix.

Obtaining a positive definite ML estimate, is, however, not guaranteed, and this requires monitoring at each step: the method above in itself does not include the nonlinear, inequality constraint of positive definiteness for the solution at each M–step.

On the other hand, finally, note that nonlinear (equality) constraints on the space of allowable covariance matrices can be accommodated: the appropriate matrix of partial derivatives replaces the matrix H of constants used above. For these problems, however, the M–step will not usually remain linear.

Example: An Order Three Toeplitz Matrix.

We calculate the fixed matrix, used repeatedly for the M–step of the algorithm, for estimation of the order three Toeplitz given in the example of §4.3. An allowable covariance matrix has the form:

$$\begin{bmatrix} a & b & c \\ b & a & b \\ c & b & a \end{bmatrix}.$$

Embedding the matrix in the space of matrices of the unconstrained form above results in nine parameters to be estimated: six associated with (the suitably replaced) symmetric U_{ij} and three associated with (the suitably replaced) skew-symmetric U_{kl}. The constraint matrix H has the form:

$$H = \begin{bmatrix} 1 & -1 & 0 & 0 & 0 & 0 & 0 & 0 & 0 \\ 0 & 1 & -1 & 0 & 0 & 0 & 0 & 0 & 0 \\ 0 & 0 & 0 & 1 & -1 & 0 & 0 & 0 & 0 \end{bmatrix}.$$

The matrix $\mathcal{U}$ has the form:

$$\mathcal{U} = \begin{bmatrix} \mathcal{U}_1 & \mathcal{U}_2 \\ (\mathcal{U}_2)^{\mathrm{T}} & \mathcal{U}_3 \end{bmatrix},$$

where

$$\mathcal{U}_1 = 2I_3 + 32J_{(3,3)}, \qquad \mathcal{U}_2 = 28J_{(3,6)}, \qquad \mathcal{U}_3 = 4I_6 + 24J_{(6,6)},$$

for $J_{(p,q)}$ a $p \times q$ matrix of all ones. The fixed matrix used in the M–step is:

$$[\mathcal{U}^{-1} - \mathcal{U}^{-1}H^T(H\mathcal{U}^{-1}H^T)^{-1}H\mathcal{U}^{-1}] =$$

$$\begin{bmatrix} V_1 & V_2 & V_2 \\ (V_2)^T & V_3 & 0 \\ (V_2)^T & 0 & V_4 \end{bmatrix},$$

for

$$V_1 = \tfrac{37}{294} J_{(3,3)}, \qquad V_2 = -\tfrac{1}{14} J_{(3,3)} \qquad \square$$
$$V_3 = \tfrac{1}{8} J_{(2,2)}, \qquad V_4 = \tfrac{1}{4}I_4.$$

4.8 Estimation in the Presence of Missing Data.

When data are missing completely at random (see Little and Rubin [1987; p. 14] for a precise definition), our proposed method can be easily extended, using a different, preliminary Jordan embedding. A simple example we believe is sufficient to make the general strategy clear.

Suppose, therefore, that the covariance matrix Σ_X is unconstrained and of order three:

$$\Sigma_X = \begin{bmatrix} a & d & f \\ d & b & e \\ f & e & c \end{bmatrix}.$$

Suppose also that the data arrives in four groups of relative completeness:

Group 1: (x_{11}, x_{12}, x_{13})
Group 2: $(x_{21}, x_{22}, \emptyset)$
Group 3: $(x_{31}, \emptyset, \emptyset)$
Group 4: $(\emptyset, \emptyset, x_{43})$

where "$\emptyset$" denotes missing observations at a component in that 3–vector.
View the data above, now, as a single sample from the r.v. W, with

$$W = (w_1, \ldots, w_7)$$

where:

$w_1 = x_{11}, \quad w_2 = x_{12}, \quad w_3 = x_{13}, \quad w_4 = x_{21}, \quad w_5 = x_{22}, \quad w_6 = x_{31}, \text{ and } \quad w_7 = x_{43}.$

Next, in the matrix for Σ_W index the parameters in Σ_X according to group membership. This is done here only to make the form of the matrix embedding clearer: $a = a_1 = a_2 = a_3 = a_4$; $b = b_1 = b_2 = \ldots$; and similarly for c, d, e, and f. We find that:

$$\Sigma_W = \begin{bmatrix} a_1 & e_1 & f_1 & 0 & 0 & 0 & 0 \\ e_1 & b_1 & d_1 & 0 & 0 & 0 & 0 \\ f_1 & d_1 & c_1 & 0 & 0 & 0 & 0 \\ 0 & 0 & 0 & a_2 & e_2 & 0 & 0 \\ 0 & 0 & 0 & e_2 & b_2 & 0 & 0 \\ 0 & 0 & 0 & 0 & 0 & a_3 & 0 \\ 0 & 0 & 0 & 0 & 0 & 0 & c_4 \end{bmatrix}$$

One checks that the following order 12 matrix is Jordan; Σ_W appears identically in the upper left corner, so the matrix provides a Jordan embedding:

$$\begin{bmatrix} a_1 & e_1 & f_1 & 0 & 0 & 0 & 0 & 0 & 0 & 0 & 0 & 0 \\ e_1 & b_1 & d_1 & 0 & 0 & 0 & 0 & 0 & 0 & 0 & 0 & 0 \\ f_1 & d_1 & c_1 & 0 & 0 & 0 & 0 & 0 & 0 & 0 & 0 & 0 \\ 0 & 0 & 0 & a_2 & e_2 & 0 & 0 & f_2 & 0 & 0 & 0 & 0 \\ 0 & 0 & 0 & e_2 & b_2 & 0 & 0 & d_2 & 0 & 0 & 0 & 0 \\ 0 & 0 & 0 & 0 & 0 & a_3 & 0 & 0 & e_3 & f_3 & 0 & 0 \\ 0 & 0 & 0 & 0 & 0 & 0 & c_4 & 0 & 0 & 0 & e_4 & f_4 \\ 0 & 0 & 0 & f_2 & d_2 & 0 & 0 & c_2 & 0 & 0 & 0 & 0 \\ 0 & 0 & 0 & 0 & 0 & e_3 & 0 & 0 & b_3 & d_3 & 0 & 0 \\ 0 & 0 & 0 & 0 & 0 & f_3 & 0 & 0 & d_3 & c_3 & 0 & 0 \\ 0 & 0 & 0 & 0 & 0 & 0 & e_4 & 0 & 0 & 0 & b_4 & d_4 \\ 0 & 0 & 0 & 0 & 0 & 0 & f_4 & 0 & 0 & 0 & d_4 & a_4 \end{bmatrix}$$

Moreover, after permuting certain rows and corresponding columns, we see that it is orthogonally similar to
$$I_4 \otimes \Sigma_W.$$
Hence the matrix is positive definite if and only if Σ_W is positive definite. Also, note that the matrix embedding just given readily generalizes to arbitrary patterns of missing data (missing completely at random).

Thus, when no linear constraints are imposed on Σ_X, the embedding above is already a Jordan embedding, and is such that it is positive definite if and only if Σ_w is positive definite, and involves no side conditions. Therefore, the original scheme of Rubin and Szatrowski [1982] can be used to give a closed form, single–step solution for the M–step in the EM estimation of Σ_w (and hence also for Σ_x).

When Σ_x is itself linearly constrained the side conditions are adjoined in the M–step of the EM algorithm as described above, but now applied to a r.v. Y obtained from the r.v. W. Thus, we have obtained, as promised, a fixed, linear M–step for the EM procedure as applied to ML estimation of a linearly patterned covariance matrix in the presence of missing data.

It is important to note that in the cases just considered, constrained or unconstrained estimation in the presence of missing values in the original r.v. X, the method may lead to embeddings that need not have a single maximum likelihood solution. Again, the standard EM caveats apply here: the EM algorithm in such cases will not have the guaranteed convergence property assumed above, and may fail to converge to any of the multiple, equal, global maxima. As mentioned earlier, the Gibbs Sampler (or other data augmentation method or one of its many hybrids) may be a good candidate for an alternative procedure.

4.9 Some Conclusions about the General Solution.

By means of an augmented EM algorithm with an essentially trivial M–step, we have solved the general ML estimation problem for covariance matrices with multivariate normal data, when the covariance matrix is assumed to have a linear structure and data may be missing. The method involves a matrix embedding of a known, fixed order, and uses a (defining) property of Jordan algebras. The linear pattern can be completely arbitrary; we applied it matrices of Toeplitz form. Moreover, given the original positive definite covariance matrix, the embedding always returns another positive definite matrix, and standard exponential family theory leads to the existence of a unique ML estimate.

We remark that many other embeddings into Jordan algebras are possible, aside from that given by Lemma 2, and the missing data example just discussed. These alternatives, also mathematically interesting in their own right, serve to highlight the non-uniqueness of Jordan embeddings, are discussed below. They are, it must be said, basically incomplete and less than satisfactory as elements of a useful statistical procedure, since either they do not clearly provide positive definite embeddings, or the order of the embedding may grow without limit.

4.10 Special Cases of the Covariance Matrix Estimation Problem: Zero Constraints.

Certain constrained matrices have Jordan embeddings that take a different form than considered so far. We look at selected examples in this section. In all cases we note, however, that these embeddings must still be considered preliminary and incomplete as solutions to the covariance matrix estimation problem because we have not (in most of these cases) been able to verify that a given covariance matrix has a positive definite embedding. We conjecture that completed proofs will be obtainable in some cases, but the main point here is that Jordan embeddings of any type are highly non-unique, and therefore offer opportunity for improved (e.g. smaller dimension, fewer additional parameters, etc.) solutions for some families of constrained matrices.

As a first example, consider the following order 8 matrix. It is a Jordan embedding of the order 4 matrix in the upper left corner:

$$\begin{bmatrix} a & h & i & j & a & -h & -i & -j \\ h & b & f & g & h & \theta_1 & f & g \\ i & f & c & e & i & f & \theta_2 & e \\ j & g & e & d & j & g & e & d \\ a & h & i & j & a & -h & -i & -j \\ -h & \theta_1 & f & g & -h & b & f & g \\ -i & f & \theta_2 & e & -i & f & c & e \\ -j & g & e & d & -j & g & e & d \end{bmatrix}$$

This is a Jordan embedding for: (i) $j = 0$; (ii) $j = i = 0$; (iii) $j = i = h = 0$. Note that case (ii) the matrix above is never Jordan if $j \neq 0$, and that for $j = i = 0$ we may fix $\theta_2 = c$, and for $j = i = h = 0$ we may fix $\theta_1 = b$, $\theta_2 = c$.

As another example, for proper embeddings of order 3 matrices we have

$$\begin{bmatrix} a & e & f \\ e & b & d \\ f & d & c \end{bmatrix}$$

embedded in the order 6 matrices:

$$\begin{bmatrix} a & e & f & a & -e & -f \\ e & b & d & e & \theta_1 & d \\ f & d & c & f & d & c \\ a & e & f & a & -e & -f \\ -e & \theta_1 & d & -e & b & d \\ -f & d & c & -f & d & c \end{bmatrix}$$

for the cases: (i) $f = 0$; (ii) $e = f = 0$.

Having displayed the order 3 and 4 cases above, we conjecture that the method is extendable to matrices of all order $r \geq 3$.

On the other hand, to further demonstrate the non-uniqueness of Jordan embeddings consider the case of a particular order 4 matrix:

$$\begin{bmatrix} a & 0 & c & d \\ 0 & a & e & f \\ c & e & a & g \\ d & f & g & a \end{bmatrix}$$

This has a proper Jordan embedding of order 8, and is also such that it does not use any additional parameters (as were introduced above):

$$\begin{bmatrix} a & 0 & c & d & -e & -f & -g & 0 \\ 0 & a & e & f & c & d & 0 & g \\ c & e & a & g & 0 & 0 & -d & f \\ d & f & g & a & 0 & 0 & -c & e \\ -e & c & 0 & 0 & a & g & f & d \\ -f & d & 0 & 0 & g & a & e & c \\ -g & 0 & -d & -c & f & e & a & 0 \\ 0 & g & f & e & d & c & 0 & a \end{bmatrix}$$

This constant–diagonal, order 4 example is one instance of how it is possible to properly Jordan embed any constant diagonal order 4 matrix, with only a single zero–constraints, into a family of order 8 matrices. In the next section we study the problem of embedding a symmetric matrix, of any order, subject only to the constraint that it have a constant diagonal (in particular, no zero constraints are allowed).

4.11   Embeddings for Constant Diagonal Symmetric Matrices.

In this section it will be convenient to denote by $Sym(m)_c$ the space of all elements of S_m that have a constant (but unknown) diagonal.

Direct verification shows that for $m = 2$ and 3, $Sym(m)_c$ has a proper Jordan embedding which is also commutative. Thus for $m = 2$:

$$\begin{bmatrix} a & b \\ b & a \end{bmatrix}$$

is already Jordan, and commutative; for $m = 3$:

$$\begin{bmatrix} a & b & c \\ b & a & d \\ c & d & a \end{bmatrix}$$

has the commutative, proper Jordan embedding of order 4:

$$\begin{bmatrix} a & b & c & d \\ b & a & d & c \\ c & d & a & b \\ d & c & b & a \end{bmatrix}.$$

We observe that this order 4 embedding is the natural extension of the first example given in RS; see, above p. 71, and also Little and Rubin [1987, p. 147].

To continue this process we require an algebraic excursion, using properties of **Galois fields**. That the argument to follow leads in all cases to commutative, proper Jordan embeddings of $\mathrm{Sym}(m)_c$ is evidently a new Jordan algebra result.

4.12   Proof of the Embedding Process for Sym(m)$_c$.

Theorem. For $\mathrm{Sym}(m)_c$, $m \geq 2$, let $n = n(m) = 2^{m-1}$. Then there exists a vector–space injection ϕ of $\mathrm{Sym}(m)_c$ into $\mathrm{Sym}(n)_c$ such that ϕ is a Jordan embedding, and such that $\phi[\mathrm{Sym}(m)_c]$ is commutative.

Proof.   We begin by pulling together some of the standard, known facts about the finite Galois fields $GF[p^k]$, for p any prime, and permutation group representations of $GF[p^k]$. Paranthetically, we remark that we found it curious that some of these standard facts seem to have been dropped from many of the modern texts covering Galois fields, abelian groups, permutation representations, etc. We are recovering here, therefore, some of this deleted technology by using Carmichael [1937], to which all page references below will refer.

The needed facts are actually few in number:

(i)   The additive group $GF[p^k]^+$ of $GF[p^k]$, for any prime p, is abelian and of type $(1, 1, \ldots , 1)$, with k 1's (p. 245);

(ii)   A permutation group matrix representation σ for $GF[p^k]^+$ is found by letting the group elements be denoted by $\{g_1, g_2, \ldots , g_n\}$, $n = p^k$, and choosing an arbritrary set of real numbers $\{a_1, a_2, \ldots , a_n\}$. Then σ is defined by: $\sigma(g_i)[a_s] = a_t$, for t = the index of $g_i + g_s$ in $\{g_1, g_2, \ldots , g_n\}$ (p. 55). We write $\sigma(g_i) = \sigma_i$, so in particular $\sigma_i(a_1) = a_i$;

(iii)   The permutation representation is a group isomorphism (p. 55), so every σ_i

$= \sigma(g_i)$ is of order p, since the k abelian generators for $GF[p^k]$ are or order p.

We specialize now to $p = 2$, and $k = m - 1$. Then:

(a)   Sym(m)$_c$ must appear identically in the upper left corner of the matrix representation for $GF[2^{m-1}]$. As proof, first observe that the diagonal of the n x n matrix (n = 2^{m-1}) is exactly a_i, since $2g_i = 0$ for all i. Next, no index in the set of a_i's, above the diagonal, in the upper left corner block of order m, can appear more than once. This must be so since otherwise the degree of the irreducible polynomial associated with the representation would be strictly less than $k = m - 1$.

In other words, Sym(m)$_c$ has been shown to be properly embeddable in

Sym$(2^{m-1})_c$,

using $2^{m-1} - m$ new, real parameters

$\{a_{m+1}, \ldots, a_n\}$ for n = 2^{m-1}.

(b)   Next, using **(iii)** above shows that every σ_i is of order 2: $\sigma_i^{-1} = \sigma_i$;

(c)   Consider now a product in $\mathfrak{R}$ of the form $\sigma_i(a_t)\sigma_j(a_t)$. Given i, j and t there exists a unique u such that $\sigma_i(a_t) = a_j(a_u)$. But then also:

$$a_u = \sigma_j^{-1}\sigma_i(a_t) = \sigma_j\sigma_i(a_t) = \sigma_i\sigma_j(a_t),$$

so that $\sigma_i(a_u) = \sigma_j(a_t)$ as well;

(d)   From the uniqueness of u given i, j and t, and **(c)** above, it now follows that:

[¤] $\sum \sigma_i(a_t)\sigma_j(a_t)$ (the being sum taken over t, with i, j fixed)

$$= \sum [\sigma_i(a_t)\sigma_j(a_u) + \sigma_i(a_u)\sigma_j(a_t)],$$

with the i, j, t and u in each bracketed sum being related as in **(c)**. Since the representation is real, and abelian by **(v)** above, we see:

$$\textbf{[¤]} \quad = \sum [\sigma_i(a_t)\sigma_j(a_u) + \sigma_j(a_u)\sigma_i(a_t)]$$

$$= \sum [\sigma_i(a_u)\sigma_j(a_u) + \sigma_i(a_u)\sigma_j(a_u)]$$

$$= 2[\sum \sigma_i(a_u)\sigma_j(a_u)] \quad \text{(the sum taken over u).}$$

We recognize the bracketed sum as the (i, j)th element of the square of the matrix for the representation. By inspection, and using **(vi)** above, we see then that letting d be such that $g_i + g_j = g_d$, leads to the conclusion that the (i, j)-th element of the product is equal to that for the (u,v)th element, given u and v such that $g_u + g_v = g_d$.

Thus, we have shown that a product of two copies of the matrix for repre-sentation based on $\{a_1, a_2, \ldots, a_n\}$ is an element of $Sym(n)_c$, for $n = 2^{m-1}$. It follows that the space of all such matrix representations, for all real n–tuples

$$\{a_1, a_2, \ldots, a_n\},$$

is a Jordan algebra.

Consider next two matrix representations of $GF[2^{m-1}]$, one specified by the real n–tuple $\{a_1, \ldots, a_n\}$, the other by

$$\{a_{1*}, \ldots, a_{n*}\}, \text{ for } n = 2^{m-1}.$$

Using the same techniques as above we can show that the resulting matrices for the two representations are commutative. Then, since any set of real, pairwise commuting symmetric matrices can be simultaneously diagonalized by an orthogonal rotation, it is clear that how our embedding procedure must work.

Thus, given a real m-tuple $\{a_1, a_2, \ldots, a_m\}$, we adjoin new a_i so as to create the n-tuple $\{a_1, a_2, \ldots, a_n\}$, for $n = 2^{m-1}$. The added constants are assumed to range over all the real p-tuples $\{a_{m+1}, a_{m+2}, \ldots, a_n\}$, for $p = 2^{m-1} - m$.

Each n–tuple so constructed generates a matrix representation for $GF[2^{m-1}]$, and the resulting space formed by all such matrices constitutes a Jordan algebra, orthogonally equivalent to a subspace of diagonal matrices of order 2^{m-1}.

This completes our proof of the Theorem. ∎

We remark that our embeddings are quite highly structured, with the expanded matrices forming, for example, Latin squares with constant main and cross diagonals. They are also symmetric about the cross diagonal and contain a row and column permuted copy of $Sym(m)_c$ in the lower right corner: see the examples below for m = 4, 5.

Furthermore, it can be seen that any of the original parameters a, b, . . . , may be put equal to each other, and the argument still goes through. Thus for covariance matrices having constant diagonal, and with linear constraints of just this type, the theorem also yields commutative Jordan embeddings.

It is interesting to observe that our argument can be extended to include the case of embedding S_m (unconstrained) into a space of commutative matrices. At the cost then of adding many new parameters, it is possible to commutatively embed any set of symmetric matrices, irrespective of any constraints on the set. We believe that the extensive structure of these embeddings might offset some of the computing penalty associated with the explosive growth of the parameter list. The embedding in question is given by:

Corollary. Given $m \geq 2$, there exists an imbedding ϕ of $\mathbf{S}_m$ in $\mathbf{S}_k$, for $k = 2^{q-1}$, $q = 3$, such that $[\phi(\mathbf{S}_m)]^2$ is commutative and so orthogonally equivalent to a subspace of diagonal matrices.

Proof. Using the facts collected above, it is immediate that $\mathbf{S}_m$, the space of real symmetric matrices with arbitrary diagonal, can be embedded in

$\qquad \mathbf{S}_n$, for $n = 3^{m-1}$.

Here, the diagonal consists of exactly the vector

$$\{a_1, \ldots, a_m, a_{m+1}, \ldots, a_n\}, \text{ for } n = 3^{m-1}.$$

No repetition of indicies can occur in this n-tuple, since then for some two elements, x and y of the field $GF[3^{m-1}]$, we would have $2x = 2y$, and $2(x - y) = 0$. However, addition takes place in a field of characteristic not 2, so this would imply $x = y$.

Next, it can be checked that this imbedding of $\mathbf{S}_m$ in $\mathbf{S}_n$ is such that the square of any element in the image has a constant diagonal, namely just $\Sigma\, a_i^2$. Hence the square is a member of $\text{Sym}(3^{m-1})_c$. It follows therefore that the square can be embedded in $\text{Sym}(k)_c$, for $k = 2^{q-1}$, $q = 3^{m-1}$, in such a way that it is orthogonally equivalent to a subspace of diagonal matrices. This completes the proof. ∎

For the reader's convenience we display the imbedding of $\text{Sym}(m)_c$ for $m = 4$ and 5. For $\text{Sym}(4)_c$:

$$\begin{bmatrix} a & b & c & d & e & f & g & h \\ b & a & e & f & c & d & h & g \\ c & e & a & g & b & h & d & f \\ d & f & g & a & h & b & c & e \\ e & c & g & h & a & g & f & d \\ f & d & h & b & g & a & e & c \\ g & h & d & c & f & e & a & b \\ h & g & f & e & d & c & b & a \end{bmatrix}$$

Skipping the letter "o" in the parameter list to avoid confusion with the number zero, we have for $\text{Sym}(5)_c$:

$$
\begin{bmatrix}
a & b & c & d & e & f & g & h & i & j & k & l & m & n & p & q \\
b & a & f & g & h & c & d & e & l & m & n & i & j & k & q & p \\
c & f & a & i & j & b & l & m & d & e & p & g & h & q & k & n \\
d & g & i & a & k & l & b & n & c & p & e & f & q & h & j & m \\
e & h & j & k & a & m & n & b & p & c & d & q & f & g & i & l \\
f & c & b & l & m & a & i & j & g & h & q & d & e & p & n & k \\
g & d & l & b & n & i & a & k & f & q & h & c & p & e & m & j \\
h & e & m & n & b & j & k & a & q & f & g & p & c & d & l & i \\
i & l & d & c & p & g & f & q & a & k & j & b & n & m & e & h \\
j & m & e & p & c & h & q & f & k & a & i & n & b & l & d & g \\
k & n & p & e & d & q & h & g & j & i & a & m & l & b & c & f \\
l & i & g & f & q & d & c & p & b & n & m & a & k & j & h & e \\
m & j & h & q & f & e & p & c & n & b & l & k & a & i & g & d \\
n & k & q & h & g & p & e & d & m & l & b & j & i & a & f & c \\
p & q & k & j & i & n & m & l & e & d & c & h & g & f & a & b \\
q & p & n & m & l & k & j & i & h & g & f & e & d & c & b & a
\end{bmatrix}
$$

4.13   The Question of Nuisance Parameters.

We conclude this work with a discussion of the "nuisance" value of the added parameters that may have been needed in constructing a proper Jordan embedding. Some initial observations follow, after which we settle down to a more rigorous look at nuisance parameters.

Evidently, the nuisance status of the added parameters is linked to both the observed data, the original parameter vector, and also to the unobserved data. And the unobserved data figures indirectly, but materially, in our iterative scheme for estimating the original parameter: the EM algorithm generates estimates of the sufficient statistics for the complete (observed and unobserved) data.

Little and Rubin [1987, p. 98] show that a linear decomposition of the log likelihood, and the factoring of the parameter space into a product space (this being an assumption, requiring verification), results in the information matrix for each subvector being found as a block matrix on the diagonal of the information matrix for the problem, given the whole parameter vector. That is, maximization of the terms of the linearly decomposed log likelihood leads to the true ML estimate for each part, and also does not change the information matrix for each part.

Also, since X is a subvector of Y, and η is a subvector of the enlarged para-meter vector ζ, say, for Y we observe that $\mathscr{I}(X; \eta)$ appears in the upper left corner of $\mathscr{I}(X; \zeta)$.

We examine the question of whether the added parameters are nuisance terms, in the sense that knowledge of their true values does not change the information about η. If the added parameters used in constructing a proper Jordan embedding can be assumed to **not** be nuisance parameters, so that they always carry, in principle, some information, then they would appear to have an inferential status more elevated than the parameters of a numerical construction: they would be only superficially similar to the undetermined Lagrange Multipliers used in the general solution to the constrained covariance problem given earlier.

One way to study the informativeness of the added parameters in our procedure is to note that, given the complete data Y, it is then well-known that the ordinary least squares estimate (= OLS, in a certain linearized form of the estimation problem: see for example Malley [1986]) for ζ based on Y is the UMVUE for ζ, while our algorithm is based on the observed data X, and still has guaranteed convergence to this OLS. This, despite the fact that for some of the data problems covered by the method, for example most variance components problems, it is known that there do not exist any (non-iterative, closed form) UMVUE's for the subvector of ζ that parameterizes the distribution for X.

Some care is required here to not make more of the remarks above than is strictly warranted, since the OLS obtained is the UMVUE for η among all estimates based on the complete data Y, which is never fully observed, even though our procedure converges to the true value of ζ that defines the covariance of Y. Thus "UMVUE" in this context refers to only a hypothetical or "virtual" data set.

Moreover, we have seen that there are many possible proper Jordan embeddings for var(X), and by implication then, many complete data vectors Y, so that an UMVUE based on one complete data vector, $Y^{<1>}$ say, may have strictly smaller variance than one based on, $Y^{<2>}$, say.

To provide some closure on the nuisance status of any added parameters, we move on to a rigorous proof of the non-nuisance character of the added parameters, using an alternative, precise definition of "nuisance" parameter: specifically, that version advanced in the differential geometric approach to statistical inference.

For this, Amari [1985] and Amari et al. [1987] serve as general introductions, while as introductions to manifolds and differential geometry we recommend Boothby [1986], and Sagle and Walde [1973]. Further, useful discussion of nuisance parameters may be found in Cox and Reid [1987]. In this last reference note in particular the discussion comments of Moolgavkar and Prentice [p. 34-35].

Working from Amari [1985; Chapter 8, especially pp. 250 et seq.] we invoke the notion of the orthothogonalized Fisher information, and (changing notation slightly) suppose the original parameters are

$\lambda = (\lambda_1, \ldots, \lambda_u)$ ($\neq$ the Lagrange multipliers), with the additional parameters being

$$\tau = (\tau_1, \ldots, \tau_v),$$

with

$$\zeta = (\zeta_1, \ldots, \zeta_u, \zeta_{u+1}, \ldots, \zeta_t) = (\lambda, \tau),$$

where

$$\tau_i = \zeta_{u+i}, \quad 1 \le i \le v, \quad t = u + v.$$

For the information matrix $\mathscr{I}(Y; \zeta)$, based on the complete data Y and the parameter $\zeta = (\lambda, \tau)$, we have:

$$\mathscr{I}(y,z) = \begin{bmatrix} \mathscr{D} & \mathscr{E} \\ \mathscr{E}^T & \mathscr{F} \end{bmatrix}.$$

for matrices $\mathscr{D}$, $\mathscr{E}$ and $\mathscr{F}$ of expectations of mixed second order partial derivatives. The **orthogonalized Fisher information** is then defined as:

$$\tilde{\mathscr{I}}(y; \zeta) = \mathscr{D} - \mathscr{E}\mathscr{F}^{-1}\mathscr{E}^T.$$

Here

$$\tilde{\mathscr{I}}(y; \zeta)$$

is intended to represent the information lost by not knowing the true value of τ, and we say τ represents a nuisance parameter if and only if

$$\tilde{\mathscr{I}}(y; \zeta) = \mathscr{D} = \mathscr{I}(y; \lambda).$$

This occurs if and only if $\mathscr{E} = 0$.

Next, consider the family of all possible Jordan embeddings for a given covariance matrix Ω.

We have seen that Ω will usually have several distinct embeddings in matrices of different orders. In the apparent absence of any obvious algebraic or geometric invariant or indexing scheme for the family of all possible embeddings, for example "matrix order + dimension," we choose to focus on embeddings with minimal dimension, i.e. an embedding in a space of matrices of some order (*any* order) such that space has minimal dimension. As one embedding may have greater order than another, but may well be computationally more efficient, our criterion of minimal dimensionality will not address this possibly important feature of the embedding process.

Assume now that Ω^* is a Jordan embedding of Ω with smallest dimension. Next, recall that the complete data Y is taken from a regular exponential family. Then, because Ω^* is Jordan it follows that the set $\{y^T G_i y\}$ is a complete minimal sufficient statistic for ζ : this importnant result is due to Seely [1972; 1977]. Moreover, from Lehmann [1986, Exercise 15, p. 66], $\mathscr{E}$ is seen to be:

$$\mathscr{E} = [\mathscr{E}_{rs}] = [\text{cov}(y^T G_r y, y^T G_s y)]; \quad 1 \le r \le u; \quad u + 1 \le s \le u + v.$$

Let $\mathscr{J} = \mathscr{J}(\Omega^*)$, the Jordan algebra in $\mathbf{S_q}$, q = u + v, spanned by Ω^*. Since Ω^* is a minimal Jordan embedding, $\mathscr{J}(\Omega^*)$ is exactly the vector space spanned by the matrices in Ω^*. Also, let $\mathscr{J} = \oplus \mathscr{J}_i$ be the ideal direct sum decomposition of $\mathscr{J}$ into its simple Jordan subalgebras $\mathscr{J}_i$. For $a \in \mathscr{J}$ we have $a = \oplus a_i$, and can define (as in Chapter 3) the **support** of a to be the set of indicies such that $a_i \ne 0$, letting the support of zero be $\varnothing$.

Using **Theorem 7.4** of Chapter 2, we have, with $a, b \in \mathscr{J}$, that $y^T a y$ and $y^T b y$ are statistically independent if and only if they have disjoint support in $\mathscr{J}$.

Suppose that $b \in \mathscr{J}$ corresponds to one of the added parameters, that is, b is the matrix associated with a quadratic form $y^T b y$ linked uniquely to one of the added parameters, where $y^T b y$ is the minimal sufficient statistic for the added parameter $\tau_{i\bullet}$, say. Let matrix b be denoted $G_{i\bullet}$.

Further, assume $a \in \mathscr{J}$ is any matrix associated with one of the quadratic forms linked with one of original parameters, where

$$a = \Sigma \lambda_j G_j ,$$

with

$$\Omega = \mathscr{S}(\{G_j\}) = \text{the space of allowable covariances}.$$

Assme that $\tau_{i\bullet}$ is a nuisance parameter. Then using the quadratic form result just quoted and the equation above for $\mathscr{E}_{rs}$, we get that $G_{i\bullet}$ must have disjoint support from all those G_j that are associated with the parameter λ. But then $G_{i\bullet}$ is not in the sum of any $\mathscr{J}_m$ such that m is in the support of any a, for a of the form $a = \Sigma \lambda_j G_j$.

On the other hand, the sum of all these $\mathscr{J}_m$ is itself a Jordan algebra, so there exists a proper subalgebra of $\mathscr{J}$ which contains all $a = \Sigma \lambda_j G_j$. However, Ω appears in the upper left corner of the space spanned by the G_i associated with all λ_i. Hence we get a contradiction to Ω^* being a Jordan embedding of minimal dimension. Thus $\mathscr{E}_{si}^*$ cannot be zero for all s, and therefore $\tau_{i\bullet}$ is never a nuisance parameter. ($\square$)

As the argument applies to all components $\tau_{i\bullet}$ of τ we get that τ is never a nuisance parameter vector if Ω^* is taken of minimal dimension.

Observe that we had earlier shown that the ML estimate for λ is just the vector of the first u components of the ML estimate for ζ and that this is the same conclusion that follows from assuming that τ is a nuisance parameter: see Amari [1985, p. 282-283]. This suggests, therefore, that being able to obtain an ML estimate by using a subvector of an ML estimate for a parameter vector with additional components is far from being inferentially equivalent to the nuisance status of the additional parameters.

Finally, we note that if the additional parameter vector τ had been declared a nuisance parameter, in virtue of $\mathcal{E} = 0$, then it can be shown that a reparameterization ζ^* of ζ is always possible, with $\zeta^* = (\lambda, \tau^*)$ such that τ^* is orthogonal to λ (see Amari [1985]). We can conclude that τ^* is basically irrelevant to inference about λ, and we remark that this result is, in fact, a basic differential geometry theorem due to Frobenius: see Boothby [1986; p. 158–164]. One proof of this uses the machinery of Lie algebras, which is the other main family of classical algebras, aside from the associative and Jordan algebras; see Chapter 1.

Thus, in closing, we find it interesting to note that the nuisance character of τ has been examined by jointly using results from three diverse technologies: a new Jordan algebra result, a classical Lie algebra fact, and the differential geometric approach to the interpretation of the purely statistical concept of a nuisance parameter.

References for Chapter 4

Aitchison, J. and Silvey, S. D. [1958]. "Maximum–likelihood estimation of parameters subject to constraint." **Annals of Math. Stat., 29.** 813–828.

Amari, S. [1985]. **Differential–Geometrical Methods in Statistics.** *Lecture Notes in Statistics*, Vol. 29. Springer–Verlag, New York.

Amari, S., Barndorff-Neilsen, 0. E., Kass, R. E., Lauritzen, S. L. and C. R. Rao, [1987]. **Differential Geometry in Statistical Inference.** IMS Lecture Notes – Monograph Series, Vol. 10. Institute of Mathematical Statistics, Hayward, CA.

Anderson, T. W. [1969]. "Statistical inference for covariance matrices with linear structure." In: **Proc. Second Internat. Symp. Multivariate Anal.,** (P. R. Krishnaiah, ed.). 55–66.

Anderson, T. W. [1970]. "Estimation of covariance matrices which are linear combinations or whose inverses are linear combinations of given matrices." In: **Essays in Probability and Statistics.** R. C. Bose et al., eds., University of North Carolina Press. 1–24.

Anderson, T. W. [1984]. **An Introduction to Multivariate Statistical Analysis.** Wiley, New York.

Bickel, P. J., and K. A. Doksum [1977]. **Mathematical Statistics: Basic Ideas and Selected Topics**. Holden-Day, San Francisco, CA.

Boothby, W. M. [1986]. **An Introduction to Differentiable Manifolds and Riemannian Geometry**. 2nd Editon. Academic Press, Orlando, FL.

Brown, L. D. [1986]. **Fundamentals of Statistical Exponential Families, with Applications to Statistical Decision Theory**. IMS Lecture Notes – Monograph Series, Vol. 9. Hayward, CA

Burg, J. P., Luenberger, D. G., and D. L. Wenger [1982]. "Estimation of structured covariance matrices." **Proc. IEEE, 70(9)**. 963–974.

Cox, D. R. and Reid, N. [1987]. "Parameter orthogonality and approximate conditional inference." **J. R. Statis. Soc., B, (45)**, 1–39.

Dembo, A. [1988]. "Bounds on the extreme eigenvalues of positive–definite Toeplitz matrices." **IEEE Trans. on Info. Theory, 34(2)**. 352–355.

Dembo, A., Mallows, C. L., and L. A. Shepp [1989]. "Embedding nonnegative definite Toeplitz matrices in nonnegative definite circulant matrices, with application to covariance estimation." **IEEE Trans. on Info. Theory, 35(6)**, 1206–1212.

Dempster, A. P., Laird, N. M. and Rubin, D. R. [1977]. "Maximum likelihood from incomplete data via the EM algorithm." **J. R. Statist. Soc., B, 39**, 1-38.

Eaton, M. L. [1983]. **Multivariate Statistics: A Vector Space Approach**. Wiley–Interscience, New York.

Fuhrmann, D. R. and M. I. Miller [1988]. "On the existence of positive–definite maximum-likelihood estimates of structured covariance matrices." **IEEE Trans. on Info. Theory, 34(4)**. 722–729.

Gelfand, A. E. and A. F. M. Smith [1990]. "Sampling–based approaches to calculating marginal densities." **J. Amer. Stat. Assoc., 85(410)**. 398–409.

Grone, R., Johnson, C. R., Sá, E. M., and H. Wolkowicz [1984]. "Positive definite completions of partial Hermitian matrices." **Linear Alg. and its App., 58**. 109–124.

Jacobson, N. [1968]. **Structure and Representations of Jordan Algebras**. Amer. Math. Soc. Coll. Pub., Vol. XXXIX. Providence, RH.

Lancaster, P. [1969]. **Theory of Matrices**. Academic Press, New York.

Lehmann, E. L. [1986]. **Testing Statistical Hypotheses**. 2nd Edition. Wiley–Interscience, New York.

Lev-Ari, H., Parker, S. R., and T. Kailath [1989]. "Multidimensional maximum entropy covariance extension." **IEEE Trans. on Info. Theory, 35(3)**. 497–508.

Little, R. J. A. and Rubin, D. B. [1987]. **Statistical Analysis with Missing Data**. Wiley, New York.

Magnus, J. R. [1988]. **Linear Structures**. Griffin and Company, London.

Malley, J. D. [1986]. **Optimal Unbiased Estimation of Variance Components**. *Lecture Notes in Statistics*, Vol. 39. Springer–Verlag, New York.

Malley, J. D. [1987]. "Subspaces and Jordan algebras of real symmetric matrices." **Algebras, Groups and Geometry, 4**, 265–289.

Rao, C. R. [1973]. **Linear Statistical Inference and its Applications**. 2nd Edition. Wiley and Sons, New York.

Rogers, G. S., and D. L. Young [1977]. "Explicit maximum likelihood estimators for certain patterned covariance matrices." **Comm. Statist.–Theor. Meth., A6(2)**, 121–133.

Rubin, D. B. and Szatrowski, T. H. [1982]. "Finding maximum likelihood estimates of patterned covariance matrices by the EM algorithm." **Biometrika, 69**, 657–660.

Sagle, A. A. and Walde, R. E. [1973]. **Introduction to Lie Groups and Lie Algebras**. Academic Press, San Diego.

Seely, J. [1971]. "Quadratic subspaces and completeness." **Annals of Math. Statist., 42**. 710–721.

Seely, J. [1972]. "Completeness for a family of multivariate normal distributions." **Annals of Math. Statist., 43**, 1644–1647.

Seely, J. [1977]. "Minimal sufficient statistics and completeness for multivariate normal families." **Sankhya, Series A, 39**, 170–185.

Srivastava, J. N. [1966]. "On testing hypotheses regarding a class of covariance structures." **Psychometrika, 31(2)**. 147–163.

Synder, D. L., and M. I. Miller [1991]. **Random Point Processes in Time and Space**. 2nd Edition. Springer–Verlag, New York.

Szatrowski, T. H. [1980]. "Necessary and sufficient conditions for explicit solutions in the multivariate normal estimation problem for patterned means and covariances." **Annals of Statist.**, **8**, 802–810.

Wilson, A. G., Malley, J. D., Pfeifer, J. C., and A. N. Petelin [1992]. "Remarks on the Gibbs Sampler and its Implementation on a Parallel Machine." In Amer. Stat. Assoc, 1992, Annual Proceedings.

Wu, C. F. [1983]. "On the convergence properties of the EM algorithm." **Annals of Statist.**, **11**. 95–103.

Index

Lecture Notes in Statistics

For information about Volumes 1 to 7
please contact Springer-Verlag

General Remarks

Lecture Notes are printed by photo-offset from the master-copy delivered in camera-ready form by the authors of monographs, resp. editors of proceedings volumes. For this purpose Springer-Verlag provides technical instructions for the preparation of manuscripts. Volume editors are requested to distribute these to all contributing authors of proceedings volumes. Some homogeneity in the presentation of the contributions in a multi-author volume is desirable.

Careful preparation of manuscripts will help keep production time short and ensure a satisfactory appearance of the finished book. The actual production of a Lecture Notes volume normally takes approximately 8 weeks.

For monograph manuscripts typed or typeset according to our instructions, Springer-Verlag can, if necessary, contribute towards the preparation costs at a fixed rate.

Authors of monographs receive 50 free copies of their book. Editors of proceedings volumes similarly receive 50 copies of the book and are responsible for redistributing these to authors etc. at their discretion. No reprints of individual contributions can be supplied. No royalty is paid on Lecture Notes volumes.

Volume authors and editors are entitled to purchase further copies of their book for their personal use at a discount of 33.3% and other Springer mathematics books at a discount of 20% directly from Springer-Verlag. Authors contributing to proceedings volumes may purchase the volume in which their article appears at a discount of 20%.

Springer-Verlag secures the copyright for each volume.

Series Editors:

Professor S. Fienberg
Department of Statistics
Carnegie Mellon University
Pittsburgh, Pennsylvania 15213
USA

Professor J. Gani
Stochastic Analysis Group, SMS
Australian National University
Canberra ACT 2601
Australia

Professor K. Krickeberg
3 Rue de L'Estrapade
75005 Paris
France

Professor I. Olkin
Department of Statistics
Stanford University
Stanford, California 94305
USA

Professor N. Wermuth
Department of Psychology
Johannes Gutenberg University
Postfach 3980
D-6500 Mainz
Germany